The
Battle
For
Existence

Philosophy, Poetry and Reality

Angelo Letizia

Printed in the United States of America

Lulu Books

2nd Edition

ISBN: 978-0-6151-6712-1

"Give me the truth
    Or give me the noose"

The Battle for Existence

I have torn apart reality; I have destroyed the very roots of everything we live in, akin to Descartes. But unlike Descartes, I do not rebuild a familiar structure (reality) with shaky foundations; rather, I leave reality in ordered, discernible and recognizable fragments. I leave it in this state for humanity to rebuild it using its inborn capacity, which I will discuss throughout the text.

When I look at the world I live in, I desperately try to grasp it. But I do not know what I am grasping. I do not know what I am truly looking at. And to quote Socrates, the unexamined life is not worth living. So, in a sense, this is my answer to Socrates. This is the examination of my life, which I have lived for 26 years. I have drifted for 26 years, never truly being able to nail my allegiance down, never able to truly believe in anything, always questioning, and always analyzing. And as much as I fight it, I have succumbed to lure of reason. It has beaten me, grounded me.

This is all for you Troy. It is not for you to believe, but for you to read, and examine. You will have to examine your own life, but as your father, I am entrusted to guide you. And this is what I give to you. For 26 years I have stared at my world in wonder, I have so desperately sought answers to questions I didn't know how to ask, and for 26 years I have remained stranded in wonder, beaten by the world's regulations, impressed by its awards, and kept repressed by its merits. But my son, you will not have the same fate. The best I can do is to give you something to accept, reject or alter. Read this with an open mind; suspend your cold hard reason, read this with your imagination. Troy, I have given you a foundation. Do with it what you will.

Contents

The Inheritance and the Manifestation of Sanity by the Majority

Genetically, human beings inherit traits from their parents. Defects and positive traits are passed down along generational lines. Reality and society are no different in this respect. Every moment of a society's existence is merely a composite of a multitude of socially and culturally inherited factors, as well as society's responses to present physical needs. Existence-as our present reality- is simply an inheritance from the previous generations' achievements and failures.

However, our inheritance (which I will discuss in detail in the next section, focusing on specific historical factors) is not a rigid, predetermined set of values or identities accepted by all. Rather, it is a varied and choppy set of ideas that are either fully or partially accepted by some members of society, while rejected by others. There are no laws governing the acceptance of past ideas. Rather, many times, their influence either consciously or unconsciously permeates society's institutions and values. Some may completely embrace new ideas, while others may remain slavishly indebted to traditional ways. "Society" is not a separate, conscious entity, rather it is a mob of individuals, each with their own self interests, which many times conflict. This mob of individuals is led by political figures, as well as guided by prominent figures such as celebrities and "pop culture gurus."

This mob of individuals, which we will call society, is in need of a guiding figure. Many theorists have pointed to a "social contract" which society accents to, in which people agree to give up some of their liberties in order to be able to live in a civilized fashion, where they then can pursue higher interests than merely self preservation. However, once this society is achieved, it then must be held in place by certain ideas. The lynchpin or crux of any society is that society's definition of sanity and order. The prominent leaders of society manifest a vision of "order" which in turn is then accepted by the majority of society as being "sane." And so, the majority of people manifest an accepted vision of sanity which must be strictly adhered to. To quote John Stuart Mill, this then becomes a "tyranny of the majority." As an example, in many societies, it is fashionable for its members to dress in accordance with certain unwritten norms. If one does not prescribe to these norms, they will then be branded as a "freak," "outcast," or loner, and at worst, insane.

While our "inheritance" is not a rigidly defined or accepted notion, neither is human experience which it is pressed upon. In fact, the acceptance of any part of the inheritance is contingent on a human ocean of minds that ebb and swell with knowledge, knowledge, which is acquired throughout the passage of time, and of which is irreversible. Such knowledge I will discuss in more detail in the following section. However, an example would be the acquisition of agricultural knowledge, referred to as the "Neolithic Revolution," which was in large part responsible for enabling humanity to become sedentary and specialize in other roles outside of food production. Or more modern example would be the creation of the atomic bomb. Both events have irreversibly changed society, whether they were both agreed with or not, each event was then "passed down" and inherited by later generations, and adapted to accordingly.

However, the one major determining factor of any inheritance in any society is obviously death. This is by no means an original idea, but has been expounded on by various philosophers. Death is the great equalizer, where the rich and poor are both eaten by worms. Some can prolong it, but no one can escape it. And it is on this tumultuous back drop that the inheritance must be maintained and advanced.

Yet, the retention of an inheritance is not solely affected by death. Rather, the entire state of natural existence is one calculated drive toward decay and disorder. A broken egg will never reconstruct itself. As said before, all living things die, but even before death, all living things are moving toward death. An apple browns in the sun, individual trees die, and while nature recycles itself, individual entities (as well as human beings) all move toward disorder. This in fact, is the second law of thermodynamics, all organisms move from a state of low entropy (disorder) to high entropy. I would even go so far as to say that existence should not be thought of "life" or a positive entity, but rather as merely an "absence of death," an absence of nothing.

Indeed, the very reason behind the creation of a government, some would argue, is to "guide" the populace to be "civilized." For without this overarching guide, society, I would argue, would degenerate into a cesspool of crime and savagery, perhaps much like the original state of nature, envisioned by Thomas Hobbes. It is not my intention to argue the inclinations of societies, I merely point out this theory to illustrate how the second law of thermodynamics may operate on a social, as well as atomic level. This is by no means original, it was proposed by the physicist Erwin Schrödinger. He saw life against the backdrop of the second law of thermodynamics.

The Battle

Within the frameworks of the social contract and manifestation of sanity by the majority, various members of society combat the inevitable approach of entropy. Various individuals and institutions over generations have tried to claim a stake and erect a lasting structure on top of the perpetual "flux" of the inevitable disorder[1]. And ultimately, this is the battle for existence. It is the battle waged by various institutions and members of society against the inevitable decay of all things. Whether it is a government, a definition of "insanity," the buying of a new car, the opening of a bank, all of these actions are blows against the chaos which lurks everywhere, which threatens our finite existence at every turn. And while the individuals who perform these actions eventually die, their labors are not totally in vain, because the institutions or actions many times outlive their creators and are inherited by the next generation. And all members of society are expected to partake in and adhere to these institutions[2].

If viewed in a larger context, this constant attempt to nail down the flux into a coherent, definable pattern is truly a battle. And what results from the labors of this battle is the irreversible gain of knowledge. As author Robert Wright claimed, information is the glue of society. I would add that information is the main weapon in the face of disorder. In humanities attempt to repel the inevitable disorder which

---

[1] As a side note I would also venture that in a state of anarchy, not all individuals would succumb to savagery, I believe that some would try to re-establish order, but again, it is not my intention to take up this argument in my present paper.
[2] For a further discussion of members who do not adhere to societies demands, see appendix A.

threatens to swallow it up, it has created (or more accurately the various individuals which it is composed of) various methods for this purpose.

I do not venture however, that each generation makes positive contributions, nor do I venture that there is a systemic or "cosmic" pattern which dictates any generations' contribution. I make my claim as a historian, by strictly observing the past; a general trend of progression seems to have taken place (albeit with much violence, disorder and confusion). It can be agreed upon that most people living in the world in 2007 AD have a higher material standard of living than in 1007 AD.[3] With this in mind, I believe that what has been illustrated by history has been a battle to progress.

If I have described a battle, that means there is something to be won by both sides. Order and chaos each had something to win. "Chaos" needs the complete submission of order and intellect by forcing it to conceive a nihilistic state with no possibilities of order. "Order," on the other hand, wants to achieve a state of imaginative simultaneity. A state of imaginative simultaneity could completely oppose the realm of chaos. But what it is imaginative simultaneity? I do not want to use the familiar cope out of "our finite minds cannot penetrate this realm." While this is probably true, I will not cower behind it. Imaginative simultaneity is a state in which all states of consciences exist simultaneously. Past, present and future, as well as all possible pasts, presents and futures all exist infinitely. Infinite infinities can exist. Perhaps some people in modern society feel this incompleteness and it manifests as mental illnesses when they attempt to define it[4]. I would not have to be addicted to the sparse, fixed and passing seconds of a joyous instant. Instead, I could savor an infinite present and a whole completeness[5]. While my language becomes esoteric, I think it is of necessity. I cannot define this state logically, so I must turn to poetry[6].

However, the conception of infinite states is not new to philosophy. It has been used many times to describe a state of paradise or heaven. However, I will take my description a step further. Instead of simply an infinite number of states impressed upon our passive intellects, I propose an infinity created by the human intellect. This infinity has already been created, but it is untamed. Indeed, man has been able to envision infinity, but only envision it, never live it. And I believe the battle hinges on this infinity. While it is untamable, it holds the key to vanquishing chaos-or order. In the state of infinity, imagination would be unlimited and boundless; it would not know the confines of personality, being, space, time, logic or language. We could invent our own fulfillment, our own satiety. We would laugh at infinity because we could bend it to some unknown end; we could use it to count and label a new existence, or we could invent a new imagination. This existence is fraught with division, division by time, space, ideas, colors, names sounds etc. Yet, would an all inclusive untied existence

---

[3] Some critics may counter this affirmation of progression with David Hume's argument of causality, accusing me of believing events will take place in a certain way just because they have done so in the past. I do not refute this criticism. I do not take the unfolding of the future to be an absolute constant. I believe that Hume's claim is a valid one, but I do not think this should cause us to cease making projections. I believe we should make predictions based on what we know.
[4] See appendix d
[5] I have heavily borrowed from Augustine's conception of reality, Boethius writings and Hegel's absolute. For a further discussion of men
[6] Herder

be the ideal of progress? Possibly. But perhaps we could imagine a divided unity, a sane madness, an infinite ocean, adrift but fixed, a happy sadness of which all experience by themselves in a chorus of one.

Infinity is a human creation. It surpasses both chaos and order and opens up an entirely new realm of experience. Perhaps infinity can somehow wed these bitter enemies. But, I do not want to simply live in some paradise. Although I am not as pessimistic as Schopenhauer when he says "if men lived in a land where turkeys flew around already roasted and lovers did have to be won, men would invent new miseries or hang themselves from boredom" I do think he brings up a valid point. The culmination of order cannot be a lazy paradise. That is a cop out. Rather it must be an imaginative paradise; it must always be actively working to imagine new existences, to vanquish the possibility of nil. Infinity must use the inheritance imprinted on it and create new, naked, unborn, unnamed things. This would be a place where poetry is used instead of logic. This is a lofty, esoteric goal, one which defies accurate explanation and leaves itself open to severe criticism. Yet, I offer only a few possibilities. Once the human intellect can harness infinity, it can create with it what it wants. I will explore this more fully toward the conclusion of the work.

But that is exactly what chaos is trying to do as well, except oppositely, chaos wants to vanquish the possibility of order. Infinity is the key to victory for both. I will discuss this concept further at the end of the paper, as well as the driving forces behind "order" and "chaos." However, in order to understand this battle, we must now explore its beginnings.

The Emergence of the Human Intellect

A question that has plagued philosophers for centuries, and indeed any inquiring mind, is how the earth was created, how did existence emerge, and was anything here before the earth? I believe that at one time before existence[7] there must have been a state of chaos. However, the term chaos must be defined. It is used in the Old Testament, and in the Koran, however it is lacking a description save for a few lines. Some common references to earth before creation are "chaos or nothingness." Some ancient Greek texts conceive of a state of confusion, where all the elements of the universe were blended together into a congealed ball of existence. The Buddhists account for this "time" period as a state of nothingness, one which eventually is disturbed by stirrings of a will. Modern scientists are no closer to providing an adequate answer. The big bang is a widely accepted theory, but the question remains, what was "here" before the big bang, how was the energy and matter needed to create the universe initially created or set into motion? My answer will no doubt please very few, either religious adherents or scientists, and it will most likely not be accepted by any. But speculation is a human beings only weapon for a task such as this one.

Before the creation of the universe believe that a state of nothingness existed. However, my conception of nothingness differs from the Buddhist and Judeo-Christian conception. This nothingness was a timeless, space-less "point." But what is called "nothingness" cannot accurately be described, because to describe it renders to it

---

[7] Indeed, time as a human intellect would conceive of it is a useless categorization in this case, this "point in time" that I am forced to use is due to the lack of adequate language to describe such a "place."

some type of quality. To say "nothing" existed, is to designate a quality, the quality of nothingness, the quality of not-being something else[8]. So what existed could not truly be "nothing[9]" because nothing cannot exist, there would be nothing to exist. This singular timeless, space-less entity could not truly be nothing simply because the possibility of other states existed. Thus the nothingness was actually an unordered chaos of non-being with the possibility of being. One of the opposite attributes of this chaos was the possibility of order, however it was not just opposite, but this possibility embodied the a state which could vanquish "nothingness" because the two were diametrically opposed. Order which we can define as a logical arrangement of matter existed in possibility by it being "not-nothing". Buddhist doctrine teaches that Nirvana is not a state a "non-being' because non-being still knows its opposite, this it is not a state of being or non-being. But they stop there, simply saying that a human mind cannot conceive of this state. I say, lets try. Even if I go insane, I want to know. Hegel said that his rational progression of teleological history began when pure being tried to conceive of itself. The opposite of pure being is pure nothingness, thus the dialect was born. My proposal is somewhat similar to these attempts.

I believe that the newly born chaos tried to expunge even the possibility of order, which existed as negative relationship. Thus, with the expunging of the possibility of order, of something, the battle between order and chaos began. Chaos is the natural result of an unexpunged nothingness. It is the "nothing-chaos' the empty set that exists so long as there are competing states to keep it alive. Chaos has expunged order to kill it, like tearing open your own flesh and removing a parasite. But the parasite still exists somewhere outside of the body, and so the body must kill it once and for all. But order refuses to die, refuses to be swarmed by the chaos. Modern science tells us that all natural things move toward a state of entropy in the 2nd law of thermodynamics[10], or a more disordered state. It is from this disordered state that order was initially expunged because it had to be killed, so nothing could truly be "nothing," with no possibility of anything else. And it is this expungment that created existence.

While I do not wish to simply usurp existing ideas or conform my ideas to them, I nonetheless see points of agreement. For one, while the big bang theory has many holes, I think part of it is compatible with my philosophy. Perhaps the "big bang" was the expungment of order. It was the violent revolution of order from chaos. To the question of what existed before the big bang, I retort with St. Augustine's cosmological

---

[8] On this point I am greatly indebted to Aristotle, Hegel, the mathematical theory of the empty set, and Buddhist doctrine.

[9] However, I am approaching this with an established (we hope) human intellect. Obviously there was not intellect to critique the "nothingness,' but the fact that the critique does exist means that the idea exists, whether I am can conceive of it or not.

[10] Here I must clarify my position. Many theories have postulated man in a battle against entropy. Yet, many of these theories see man, while creating local order for himself, doing this at the expense of universal order, which diminishes. I do not have the credibility to argue with a physicist, yet, I am not concerned with universal and local entropy. I believe man, and any animate creature, is n a constant struggle with entropy. And while on a scientific level he may generate more entropy by battling it, what else is he supposed to do? I believe order is in a perpetual struggle and this increased entropy is of little consequence to it.

theories. He states that is erroneous to conceive of their being "time" before" the universe was created. "Time" emerged after.

I want to expand on the notions of time and space. To echo Immanuel Kant, time and space should not be thought of as inherent in the universe, but merely as predicates of appearance. I venture to alter this view slightly. After the expungment of "order," order immediately sought to survive. Order spawned the concepts of time, space and matter to combat nothingness. While order spawned these manifestations as weapons, they nonetheless succumbed to chaos-nothingness. Chaos-nothingness, trying to vanquish order, spawned its opposite, entropy or disorder. While time, space and matter are derivations and weapons of order, they nonetheless are susceptible to entropy, and could not become completely successful bulwarks of order. Matter, while it is an impenetrable substance, while it is "something" as opposed to nothing, it is also finite, it ends, it will succumb to disorder, it is not permanent. Space, while it is able to be filled by matter, separates it. It cleaves the distance between objects and separates everything. Time, while it gives matter a reference and placement among successive states of space, time nonetheless isolates all matter and space into one solitary moment, unconnected to any other moment save for the relationship of succession. But no two moments can ever be truly connected in a coherent and sensible totality.

Despite their shortcomings, Space, matter and time became the first state of order. While it is hard to conceive, this was an imperceptible state, because there was no living consciousness to perceive it[11]. Yet, order, every evolving, ever trying to struggle against disorder, evolved into a more sophisticated form. Inanimate matter eventually spawned living matter, which, (I borrow Herders theory here) was merely a higher organization of matter into a state of consciousness, which was able to perceive of its self. Living matter, organized into a higher state able to perceive of itself[12], eventually evolved into the human intellect. Herder theorized that matter was a unitary substance, and at its highest level of organization, was mind. I believe this theory holds in part today, in light of modern neuroscience. The brain, which is the seat of the mind, is matter. Various chemicals, nerves and electric impulses make up the components of the brain, and consequently, the mind. The brain contains the same "stuff" as the rest of the body, but is organized differently and thus, is conscious, and fits into my conception of "living matter."

However the first perceptible instance of the presence of order was in the form of self-preservation which was a distinctive feature of the aforementioned conscience and perception, or the higher organized matter. Self preservation or the natural survival instinct in any life form[13] tried to preserve that beings existence, or its ordered cohesion of molecules, for as long as is possible. However, at best, self-preservation can only achieve a temporary stasis. Eventually, all things die. Yet, modern science tells us that for billions of years, simple creatures existed on the earth, creatures that simply survived for a short time, reproduced, and then died. Gradually, these singled celled simple creatures evolved into higher and higher forms of life. Evolution is an

---

[11] At this point I must add that we can not try to perceive of the expungment of order occurring in time or space, because those ideas would follow. I am using Augustine's argument to bolster my point.

[12] I am specifically referring to all animal life forms, from single celled organism, to insects, to gorillas.

[13] Human beings do not necessarily adhere to the survival instinct. This is further discussed in Appendix A.

aspect of the battle of chaos versus order. It is the fighting of disorder; it enables an inheritance to be transmitted to a new generation, it is another weapon of order.

Once order was expunged, it was not strong enough to survive. However, while a singular state of order was not strong enough to wage battle against disorder alone, what it could do was imprint its self to exist in another existence. As ordered pieces of nature evolved, they imprinted forms which would enable their succeeding forms to survive longer and be better equipped to fight disorder. I believe this is the definition of natural selection. The process of evolution began after the expungment of order from chaos. Order, now freed, began to accumulate existence to preserve itself. Finally, after billions of years of simple preservation and survival, order eventually evolved from mere preservation to intellect. Human intellect, in its primitive and simple form, differed from mere preservation because intellect could conceive of time, time as a past (which could be learned from), as a present and as a future (which could be prepared for).

I must stop here and address a point in which was brought to my attention. In my discussion of matter, I do not mean to imply an "intelligent design" theory, in which matter evolves to a divine or inherent design. All I propose rather, is that matter (as a bulwark against entropy), simply tries to battle chaos by constantly trying to overcome it, by constantly trying to better itself into forms more able to fight, like two opposing magnetic poles. Man may be an accident, but man still has a purpose, he is not completely aimless. Man is one of many possible bulwarks that could have been created against chaos.

As to the "driving force" of matter, some may postulate that I have given an irrational power to inanimate nature before living matter evolved. But even today, does not inorganic, inanimate matter and occurrences have a sort of power? Consider wind patterns, ocean currents, tectonic plate movements, earthquakes, gravity, planet rotations, atmospheric pressure and any natural force in the universe today. This "inorganic power" of nature was active before matter became organic (and still is). It is not that difficult to imagine inorganic and inanimate matter moving according to "its" own pattern. And while my writing of this force sometimes appears to give it a personality or consciousness of sorts, this is misleading. I believe the "expungment" is akin to two north magnets. They repel each other because of the way they are structured. There is no consciences element, but nonetheless, they repeal each other vigorously. Another example of the unconsciousness expungment is of a human body trying to rid itself of a virus. Yet, this analogy must be reversed because the universe, unlike a body, is trying to expel the harbinger or order, not chaos (which to the body is a germ or virus). In this light, humans can be viewed as a reverse germ being coughed up from the universal chaos. Once this germ is completely expelled, chaos can return to its true from, which is nothingness.

The human being (no longer an animal) could now conceive of his past, he could differentiate between objects in his universe, and soon he began to speak and communicate. Language was a further evolvement of intellect because preserved accumulated knowledge for later generations. Also with intellect, man began to love his kin. Love and altruism are highly controversial terms, but nonetheless, they are ideas that can transcend one self. When one person will risk his or her own life for another, when one person denies their instinct of self-preservation, love transcends the self. Men began to travel in packs and cooperate with one another. Cooperation is similar to the idea of love in the fact that it deals with the transcending of the self[14]. In the absence of written records, we cannot accurately know the truth, but what we do know is that at one point, intelligence did evolve. Whenever this juncture was, it was crucial. For at this point, the battle assumed a brand new facet. Reason was introduced.

However, a further explanation of intellect is needed. What is this faculty of reason that had spawned from the brain of man? Was it an inevitable occurrence? Was it a standard in every human brain from the moment it appeared? Was (and is it) conditioned by culture, language and experience[15]? I will discuss some of definitions, aspects and limits of reason later on in the paper. However, I believe the intellect that was born of primitive man was the ability to create, namely to accurately speculate, to envision situations, and then to go on and create those situations. By building tools, by cooking his food, by farming and storing food, by domesticating animals, by trading with other men, man had envisioned something before he actually saw it, or before it became empirically true. Man made it empirically true. Man willed his own surroundings; he tamed a part of nature and was no longer a passive spectator. This ability to create instantly became an impasse in the battle. By using his intellect, man elevated himself from the domain of animals who merely survive without any conscience knowledge of their struggle to a conscience participant in existence. He shattered the continuum of his animal world and named the fragments accordingly.

When Homo Habilis first used a rock to smash the skull of a fresh kill, he was not simply bashing a skull. Although he did not know at the time, by using forethought, by using reason, he had smashed disorder, he staked down a small piece of the flux and began to give it some order, he began to count it and categorize it. Eventually, his intelligence would enable him to fasten the various pieces of the universe into a coherent fashion, into a life, into a society, which he could pass on to his children. Our primal ancestors had led the charge and won an impressive victory against entropy. They began to think. Humans could begin to work together, instead of competing with each other, as in nature.[16] A vast new realm of possibility had been opened, or more like a black hole had spawned from the chaos, an all encompassing vortex that had the ability to reason, to order the chaos, and more importantly, to create new chaos's.

From the inception of the human intellect, the battle became pitched. It resonated throughout existence, it reached back to eternity through the past and future, because man now had the ability to write his history and think ahead. Vast new dimensions were torn open like tender skin and the intellect drank the blood of reality like some holy vampire. Reason became a usurper; it malignantly spread from its primitive origins, to tyrannical reign. First it built tools, and then it built civilizations.

---

[14] However, cooperation differs from love in the respect that men cooperate with each other in order to gain something for themselves, while they love each other (at least in its purest sense) for no benefit at all.

[15] Beiser

[16] Indeed, I do not suggest that man, once acquiring his intellect, instantly worked together. Nature is a state of blind preservation, with atomized entities competing with each other. In the realm of nature, order is fragmented and primitive, but nonetheless, it exists. And it was from this primal state that higher states emerged, namely intellect. Men still compete with each for their own versions of progress and order, and I believe this to be one of the biggest detriments to the realization of order.

Through blood stained teeth the intellect laughed, it began to build on the entropy. After a while however, the intellect forgot its humble roots and ignored its foundation, it ignored the fact that all was "built on a fairies wing[17]." Human beings possessed the intellect, and they cooperated together and began to farm and build little villages which turned into cities. They created governments and religions, they learned the laws of the universe (or at least some interpretation of those laws) and they eventually built the atomic bomb.

And yet, this explanation views existence as Manichean, when in fact, existence is merely the combination of intellects. Obviously, not all people adhere to "progress and order." The existence of crime and violence obviously speak of a determined effort to stem order. Yet, what if violence, perpetrated by an intellect, whether be a petty thief or Hitler, is merely that specific intellect's method to battle chaos? Violence, bloodshed and holocausts are maybe flawed attempts to stem the tide of entropy. However, violence is a form of chaos because of its regressive nature, but it can be manipulated to pursue order, such would be the case of a revolution that accomplishes a goal. Violence is chaos, and cannot be a means in itself, but has been used to achieve order, but usually results in more chaos. Most violence leads further into chaos, whether it be someone who snatches a purse, or whether it be an ethnically cleansed Bosnian Muslim.

However, this lumping of violence into the category of flawed order may be incomplete. Instead, I think a division must be made into the types of violence-and more generally-the types of ways in which people try to stem order. I believe that a petty thief, a rapist, a murderer, a suicide victim all have capitulated to chaos. They somehow feel the swell of chaos, it beckons to them and they cower before the overwhelming power of *nil*. Perhaps their lives seem to be one monolithic continuum of pain, one static mural of dreary colors. The only way to stop the pain of nil is to put a gun in your mouth, or rape someone who does not feel it. While criminals obviously have different motivations for committing crimes, I do believe many are cowards, and take the easy way out. However, I am not refuting my earlier claim. I think some types of violence, maybe violence on a larger scale, such as political violence, religious violence and genocide, are flawed attempts to conceive of order, or the attempt of ones to force their flawed or perverted order onto others. Now, this is easy for me to pen from my cozy suburban house with a well-stocked refrigerator. There are different factors that make a person commit a crime. However, the type of crime committed can be an indicator of a deeper motivation. If a man robs a drug store to steal food for his sick child that he can't afford, I do not think he is a coward, or has submitted to chaos. Rather he is trying to preserve his daughter. However, a rapist or child molester, either succumbs to chaos as mentioned before, or is deluded and believes that he has found order, but it is an illusionary order which is not permanent and is founded on the principle of stripping an innocent person of their order.

Yet, this begs another question. If order does triumph, will every individual partake in the victory? Unfortunately, I do not think this to be a feasible option. As mentioned above, some people adhere to chaos, simply because it is easy. It is easy to not work toward order; it is easy to not contribute to society, it takes effort, an effort that some people refuse to make. And in the absence of their effort, other members

---

[17] Fitzgerald

must pick up the slack (i.e. police offers). To paraphrase the author Jack London, some people do not live, but merely survive. And it is true, many people either waste their entire existences chasing material possessions which should be means, not ends. They try to gratify insatiable sexual lusts. And when they die, when their material existence is used up like burnt paper, what do they have to show? What do they imprint on the next generation? What knowledge do they pass down? There actions have "forked no lightning." They waste the time of productive members of society. But alas, I believe this situation to be unchangeable. This is not an absolute law, just a fact of experience. I could be wrong and hopefully I am. But I believe there will always be a sizable portion of the population who adhere to chaos due to its easy accessibility and the lack of effort needed to pursue it. My consternation at this segment of society is perhaps evolutionary inherited as Robert Wright suggests, because in their slothfulness and laziness, I realize that there will be more work to be done, that I will have to pick up their slack in this group effort that one labels as "existence."

I will discuss the intellect's evolution through the various historical stages. At each stage (and the stages are not predetermined, but only accessible after the fact) I will discuss how the intellect waged war with chaos. From the inception of the intellect and the heightened position of the battle, the intellect tried to subjugate disorder. It broke the jaw of its adversary and believed it reigned supreme in existence, but as history has proven, almost all previous attempts to subdue it were exposed.

And so, from the dawning of the human intellect, a conscience progress fought through the entropy that enveloped it on all sides. It is here where humanities battle began. This impasse was undoubtedly the crux of the battle, the rampart in which progress hinged its ultimate victory. But it was not a complete victory. Entropy may have been subdued, but it was not eradicated.

Thus, this is my answer to the age old question of "why something instead of nothing." The "something" is a result of the "nothing-chaos" trying to expunge all possibility of existence. But once order was expunged, it began to wage a battle, a battle for supremacy. I do not know which will win, I do not put fort any method, teleos or divine sanction to predict the winner. All I can state with certainty is that it is a battle because order or "something" is an aberration from the true nothing which is the natural state of all things. But true nothing cannot exist with the possibilities of order in its non-being. The only way for nothing to truthfully be nothing is to vanquish the possibility of order, of something. Thus, the only way nothing can exist with these lingering possibilities is in a state which I have labeled as "chaos." This chaos battles reason, trying to vanquish it, ultimately trying to revert to the true state, the true "nothingness" with no possibilities. Every piece of nature, organic and inanimate, is fighting this battle every second of its existence. Every tree, blade of grass and newborn baby are trying to stay alive, to preserve themselves in the face of death, or dissolution of its molecules back into to the disordered state which cannot truly become nothing because order has waged a battle for its survival. What is achieved is a sickly stasis, one of order pulling away from disorder, trying to preserve itself, which it does, but only for a short time. When it does eventually die however, the possibility of order still exists, thus the battle still rages.

However, the charge may be made against me that "your philosophy is not truly chaos versus progress, but chaos verses progress verses nothingness." I believe Chaos to be a *form* of nothingness, a state in which possibilities of other states exist. However, chaos can only be the *natural* state of things, so long as the possibility or order exists. While these possibilities exist, chaos is the closet form to true nothingness. The expungment of order was necessary to vanquish it, but in a paradoxical way, nothingness, while needing to expunge order to vanquish it, at the same time must also must rely on order to produce a state of consciousness that can conceive of true nothingness with no possibilities, thus creating a true state of nothingness. However, I do not put forth a relationship of a "tension of opposite" or chaos needing what it is fighting to exist. Simply, order and chaos are diametrically opposed; they are battling with each other for supremacy. And I believe the only way for chaos to finally vanquish order is to force it to conceive of a higher state of consciousness, one lacking any possibility of order. I will discuss this later in my paper.

The Battle throughout History

Throughout history this battle has been showcased in various forms and embodiments. The Egyptians most revered god was Osiris. Osiris was a mythical king of Egypt. He treated his subjects with compassion and was revered by all. He gave the knowledge of farming to the people. However, his brother Seth envied him and wanted to be king. So Seth murdered Osiris, and then hacked his body to pieces and spread them over the universe. Yet Osiris's widow, Isis, collected all the pieces of her husband, and with help of Anubis, resurrected Osiris body long enough to impregnate her. Osiris then went to rule the underworld, being its first citizen. To the Egyptians, Osiris was the embodiment of order. They prayed to him to see that their lives had order. Seth was the god of disorder, of violent storms and pure chaos. Seth had to be appeased with offerings to be kept at bay. His violence and sexual lust knew no bounds. And so the Egyptians world was a battle ground for the two brothers, for chaos and order. The resurrection of chaos rebutted the tyrannical reign of disorder, much like the formation of the human intellect.

Cain and Abel are another familiar story, Cain, the murderer, the pastoralist, the jealous sibling who clubbed his brother to death. When questioned, he answered God with "am I my brothers keeper?" He bludgeoned Abel, he bludgeoned reason to death. While these two allegories are lacking in specific instances for application to the present concern with the battle for existence, especially in light of naturalistic and scientific truths, these allegories still serve as an illustration of our ancestors' acknowledgment of the battle that faces all of humanity, indeed which rages in every atom and instance of existence.

However, the impasse in which the battle reached a fevered pitch was in nineteenth century Germany. Two philosopher's theories squared off like Roman gladiators. While they were greatly indebted to their predecessors, G.W.F. Hegel and

Arthur Schopenhauer outlined two vastly opposing paradigms. Hegel tried to bridge the gap between the noumenal and phenomenal world left by Immanuel Kant. Hegel envisioned reality and history as a monistic organism powered by passion and reason, progressing ever forward toward an absolute state, a state in which the discriminations between subject and object, between faith and reason vanish-what he termed "the absolute." Thus all of history was a rationalistic progression toward a divinized state of existence in which reason reined supreme, in which the "I became the we, and we the I, in which the mind (the subjective or phenomenal) recognizes itself in the external (objective or noumenal) world.

One of his contemporaries, Arthur Schopenhauer, violently rejected Hegel's assessment of reality. Instead of reality being a rationalistic progression, Schopenhauer envisioned reality as nothing more than the appendage to a blind irrational will. The "will" was the all pervading force; it was Kant's "thing-in-itself," the realm behind appearances. Reality was absurd, not rational. Existence was nothing more than the blind stirrings of this space less, timeless, causeless, irrational will.

What interests us in our assessment of the battle are not the nuisances of either of these philosophers' theories, but what these theories represent at a specific point in time. At that juncture in time, after the effects of the French Revolution and the Napoleonic wars, the beginnings of industrialization and the scientific revolution, after the influential theories of Immanuel Kant, Hegel and Schopenhauer unconsciously illustrated the battle in contemporary society. There theories exposed the bare naked fact of our existence by bringing us face to face with the battle for it. However, at the time, there were still lingering vestiges cloaking the battle, which I will discuss in the next section. According to the historian E.P. Thompson, a historian can only accurately access history when it is past and he reflects back on it. That is precisely what my study attempts to do; it aims to dissect an instant in history, a crucial juncture, a juncture that exposed the battle for existence.

I recently have had the opportunity to witness this eternal battle first hand. I watched the birth of my son. I watched as space, matter, time and intellect congealed into a flesh, I watched him rip himself from chaos and fight to be born. I watched order, in the form of my son, hack through bone, blood and entropy to be part of this world, to be order incarnate, to be a tiny bulwark against entropy. Outside, the trees twisted from the earth like gnarled hands, desperately trying to choke the entropy from their own cells. But the trees will die, as will the ground they spawned from. But my son will one day be able to imagine the infinite.

"The Failure of History" A Historical Analysis

In my discussion of history I do not want to pursue a Hegelian vein of thought which all of history follows a dictated, premeditated or divine pattern. Despite a somewhat similar discussion of the "philosophy of history," I aim simply to study history as a historian, not attributing to it hidden meanings or symbols. However, I do intend to view history in a practical, as well as conceptual way. Hegel's vision of history is fueled by the dialect, the rational progress of incomplete matter realizing its deficiencies and synthesizing them into a new whole.

Recently, another author has put forward similar views to mine. However, there is a crucial difference. Robert Wright, in his amazing work entitled "Nonzero: The Logic of Human Destiny," stated that essentially, all of history was a creative thrust. I only partially agree with this notion. Instead, I believe one can only make the assertion in retrospect. History is a battle. My hypothesis pits two competing and dynamically opposed forces against one another. I do not proclaim any spiritual or abstract dialect. Rather, I would argue with telling evidence, the battle in which I speak of is present in every particle of nature. Whether it is a calculated conscious effort to stem the tide of disorder by a complex civilization, or a stray cat that maintains a degree of self-preservation by eating a dead bug, the battle is an observable, tangible event that rages in every being and entity in nature. The battle is the atom of all existence, the origin of existence, from whence all existence ensues. The battle rages in every living thing, it is the all pervasive facet of life on this planet, of life in this universe, it dictates every moment, every movement, every action of history. In a sense, history is the record of man's battle against entropy. However, it is also the record of his failures and his contradictory inheritances resulting from the battle. Thus, I term it "the failure of history."

What I intend to illustrate is the failure of various attempts to combat entropy; empire, religion, humanism, reason, science, nationalism and materialism[18]. Yet, on the same token, these defenses were only failures in the sense they did not lead order to victory, they are not ends in themselves. Yet, in their failure, they became means to an end; they became vessels for victory-if used correctly.

History is philosophy in action, it is thought in action. Philosophy is an abstraction without the blood of people, without the voice of a dying man or the cries of a newborn. By studying history, my philosophy will be more clearly illustrated, it will become humanistic, not a hopeful illusion. Once again, I do not believe that history can be tailored to fit an idea. I will present different time periods in history, as well as different cultures, but I do not suggest that everyone living in a specific time period have the same beliefs, or that ideas be contiguous across cultures and classes. Rather, I aim to present a general consensus, to try and pinpoint the manifestation of the majority. Once an idea is presented into society, it is difficult to repress it. Once an idea is presented into history, even if it is vehemently opposed, it nonetheless can be come part of a generations' "inheritance." The idea can be transmitted and absorbed into a society's framework, even if a majority or plurality of people disagree or remain ignorant. In my study of various historical periods and ideas, I want to examine some of these ideas and how they influenced contemporary society and how the succeeding generations inherited them.

However, a crucial and often overlooked point must be examined at this juncture. Before we divulge into different historical inheritances, we must first ascertain which history we are looking at. Most western philosophers have only considered the importance of western history, and even more misleadingly, have assumed that only western history is of importance. Hegel discusses China, India (and dismisses Africa

---

[18] These are easy references, and some may citize me for a nominalist perspective. I will address this later in the work.

as primitive), but he subordinates their histories in a teleological progressive which ends in a superior western vision.

I cannot find any justification in assuming that western history is the pinnacle or the sole repository of truth in existence. There are billions of people in this world who do not follow Jesus Christ. While I myself am a Christian (Catholic), I think it is extremely prudent to exercise tolerance when discussing history. As Jacobi believed, philosophy-and ultimately reason-are just products of any given age and not objective standards of truth. While the west has assumed the dominant role in world affairs in the past 500 years, this gives some credence to examining western history in detail, but the histories of China, India and Africa cannot just be ignored or regulated to "heathen histories." I will cite specific instances in western history of the battle between progress and chaos, but I will also try to examine episodes in other regions.

With this said, I have come to the historical heart of my analysis. Almost as soon as intellect waged its conscious battle against entropy (alongside nature's blind self-preservation) the battle took on a brand new facet. As intellect (intellect not as an independent entity, but rather as a combination of individuals) began to conceive of its plight against the tide of disorder, it began to assemble methods in which to easier combat this disorder. The intellect evolved ways in which to control the disorder, namely, government, religion, science, industry and nationalism. These methods obviously varied and as I stated before, by no means were all encompassing of all members of a given society, they did not and do follow any cosmic pattern. But these defenses against entropy did serve as a bulwark for a majority of many of the societies I will discuss, or at least for an influential and sizeable minority. Another crucial aspect to these "defense mechanisms" is that almost all of them overlap in any given society, and many worked in conjunction with each other or serve to encourage one another.

While these methods were not contiguous or all encompassing, and were by no means accepted by all members of any given society, we can nonetheless assume their importance and role in that given society. A combination of individual intellects (usually the leaders of society, such as government officials) becomes like an oligarchy and manifests its sanity to a willing public; various members of the public then embrace this vision, and manifest the ideas back onto themselves. However, this section will attempt to dissect what actually was manifested at different junctures in history, not simply historical abstractions or symbols, what manifestations were believed and lived by the majority of people in that given society will be the focus.

Yet, ultimately, all the bulwarks that will be discussed have failed. Not in the sense that they are obsolete, in fact almost all are valid, some are indispensable. However, all of these bulwarks are compromised because each can (and do) limit the realization of the infinite within the human intellect. The infinite is blunted and subordinated to fit into a particular system. The systems are means however, and not end in themselves.

The concept of government is crucial to our discussion. I am not so much interested in the different forms of government and what they entail, as I am interested in what the existence of government means. I use the term of government very liberally, and I intend it to cover a wide variety of definitions. For our discussion, "government" is simply the form of order kept by human beings over other subordinate human beings. In Ronald Wright's work, "Nonzero," he labels the beginnings of government and social structure using the "big man theory," which states that in early tribes, one person or a group of people had control of a majority of the resources for the tribe. Usually, those in control could physically dominate the others. And so a small group of people assumed control of the resources of tribe. The "Big Man" or "Big Men" of the tribe then distributed these resources accordingly. Norman Cantor, in his work "Civilization of the Middle Ages," points to land owner ship as the hallmark of control in early Mesopotamian society. This model was then transmitted from Egypt into Europe.

I believe both of these points are valid ones and I think they mesh with my field of study. Truly, what is the purpose of the creation of a government? Why would human beings not live in a state of anarchy, similar to the animals, where it is everyone for himself? At most, how and why did human beings ever progress to create complex governments? While the aim of my study is not a discussion on the evolvement of government, I do believe that governments, whether tribal structures or massive bureaucracies, gave people a major weapon against entropy[19]. If all living structures eventually move toward entropy, chaos and disorder, the creation of a stable government would allow human beings to preserve more knowledge and enable them to pass down this knowledge to their kin in an easier fashion. Government was a major victory for order, because organized governments are directly opposed to chaos because their purpose is to keep order[20]. And so, I believe the concept of government evolved as a defense mechanism against disorder.

Once again it is not my intention to discuss the evolution of governments. However throughout history, there have been a variety of different governments throughout the world. From these early tribal beginnings, the most common form of government that evolved was along the lines of dictatorships, where one person (with the help of others) controls the rest of society. The culmination of centralized government would undoubtedly be the form of empire. It is a total incorporation of various constituent parts, under the control of a totalitarian regime. Various regions and peoples are incorporated under large empires. The empire is the logical outgrowth of governmental power. I do not suggest that all powerful governments inevitably became empires. All I suggest is that the form of empire is an all encompassing government which is the pinnacle of control. It is a massive coordination of individuals to establish a lasting bulwark against disorder and chaos. Governments vary from time a place, and largely depend on individual circumstances, but the form of empire is the culmination of government due to its authoritarian and controlling nature, which is the underlying purpose of government.

Throughout various regions and times in world history, empires have been employed to rule over territories. Various empires have been created and maintained throughout history in different regions of the world. Egypt, Mesopotamia and China all had empires, however these empires were small in comparison and did not control many territories. However, later empires would consolidate over thousands of miles of

---

[19] Wright
[20] However, in many cases, governments sometimes exacerbate chaos for a certain group of its population, i.e. 3rd Reich.

territory and millions of people under one central bureaucracy. Most notable of these later empires were Persia, Alexander's Hellenistic Empire, Rome and Mongolia.

The Persians are recognized as the world's first major bureaucracy. From humble beginnings, The Persians conquered various groups in what is today modern day Iran under Cyrus the Great. Later, the Persians consolidated their rule over such various regions as Greece, Phoenicia, Mesopotamia, Egypt and Israel. Under the Persians, what can be termed as a "global network," emerged. Various groups were put into contact with each other under the auspices of one central government. The Mongols formed a much larger empire in what is today Russia and China. Like the Persians, the Mongols controlled the lives of millions of people and formed a similar global network, assimilating Chinese, Russians and other Asians into a coherent bureaucracy.

However, the mightiest of these empires would undoubtedly be the Roman Empire, which, like the Persians and the Mongols, and even Alexander's empire, covered thousands of miles of territory and ruled over millions of people, but what the Roman's possessed that these other empires did not was longevity. The Roman Empire lasted roughly from 29 BC until 500 AD. Preceding the empire from 509 BC to 29 BC was the Roman Republic. So, for almost a thousand years, a central Roman bureaucracy ruled over lands from Spain, to England, to Egypt to Mesopotamia.

In reality, the role of an empire is to give order to society, to structure it in a way that forms networks among its citizens, a method for these citizens to rise above mere survival and achieve lasting greatness. While many citizens suffered from abject poverty in all these empires, (and this poverty was one of the contributing factors to the fall of Rome), nonetheless these empires offered a way for many citizens to pursue their desires and elevate their status. In a conceptual framework, these empires offered more than just the lures of riches to some, on an abstract level; the empire is a bulwark against anarchy, against entropy and disorder. Even poorer citizens could embrace the laws of the empire, embrace the fact that they belonged to something. These empires were various intellects attempt to nail down the flux and create a lasting achievement. Indeed, governments did not start with empires, forms of governments emerged probably in pre-historic times, hierarchal and social structure emerged in early clans. But the empire, in all its forms, can almost be viewed as a culmination of government, because of its all encompassing range, over space, population and time.

Many studies have been done on the Greek Polis and the Sumerian city-states, and these are all examples of defenses against entropy. Any governmental structure is. However, I have chosen to highlight the form of empire because of how many people are affected by it. But, it is a form which has not persisted to the present day. There are no empires in the world in 2007. Why is this? Why has the form of empire failed as a defense against entropy?

There is no easy answer to this question. I will argue that the bureaucratic structure alienated much of the population, especially economically. Specifically in the Roman Empire, the vast amount of wealth concentrated at the top of society remained frozen while the majority of people began to starve. In a time period where the standard of living was extremely low, this spelled disaster for the poorer segments of the population.

However, I believe the crucial factor in maintaining an empire is its use of force.[21] In every empire that I have mentioned, a considerable degree of force was used. Obviously, to conquer another area, a group must use force or make the threat of force apparent. Despite being lenient to conquered subjects, Rome, Alexander's Hellenistic Empire, Persia and Mongolia all used a degree of force to control and subdue its subjects. This use of force in the modern world is now unacceptable due to the destructive capability of new military technology, mainly nuclear and chemical weapons. Imagine if various groups sought to out right conquer other groups (and this does occur in limited and contained spheres of the world) with modern military technology. The events of September 11th shocked the nation, and forced America to reconsider many of its policies. The world would undoubtedly fall into the very anarchy that governments were trying to fight if empires still existed today[22].

As well as the threat of annihilation, differing attitudes concerning humanity are also present. An idea that will be discussed in more detail later is one of humanism, or the belief of human potential. Many institutions today hold human life in a very high regard.[23] The use of force and the action of forcibly subduing weaker people is not one that is universally accepted as a natural part of life anymore[24]. One great example of this is the Gladiator fights that took place during the Roman Empire. Slaves, criminals and anyone who was deemed expendable were thrown into the arena to fight to death, or to be slaughtered by beasts. Thousands flocked to the coliseum to witness this bloody spectacle. The violent extermination of expendable people was commonplace in daily life for the Romans. If I were to equip two homeless men with swords and tell them to fight to death, I would be arrested and reviled by many. Even animal cruelty is punishable by jail time. Whereas in ancient Rome, violence was an accepted and expected part of society.

The Persian Empire did not have these bloody spectacles, but nonetheless, one reads in Herodotus of how the Persians most revered rulers brutally attacked and subdued other groups and incorporated them in the empire, despite treating many of them with tolerance. The Jews label the Persians and Cyrus in particular as "the great liberator." However, the very failure of the Persian Empire can be traced to Darius III- the man who lost an empire-and his defeat at the hands of Alexander. The Persians lost their empire due mainly to their lack of force, as did the Romans when they were unable to repulse the Barbarians[25]. Every empire eventually fell, and every empire that did fall could no longer maintain the amount of force necessary to hold them together. Even modern empires have fallen prey to their reliance on force. Napoleon, the Ottomans and even Hitler were incapable of maintaining the degree of force

---

[21] I am greatly indebted Hegel on this point. He state that the Roman Empire was built upon force. I am merely extending this idea to other empires, and using this as a reason for their infeasibility in the modern world.

[22] I have to agree with Frantz Fanon on this point. Fanon said that the artillery shelling and scorched earth policies are obsolete in favor of economic dependency and sanctions.

[23] Some will also disagree with me on this point as well. I cannot say that all institutions or governments hold human life in high regard, but overall, the value of human life has increased considerably in the last 500 years. And if one institution and/or government does abuses civil rights, they usually have pressure put on them by the world community. Whether this pressure from the world community amounts to legislation for human rights is another complex issue which I will not attempt to undertake.

[24] However, there still are many governments that rely on force and coercion to attain their ends.

[25] However, the Romans, like all other great empires, "rotted" from within, and then bowed to external pressure.

necessary to sustain their empires for whatever reason. And so, while the various intellects have proposed empire as a solution to defeating the ever encroaching entropy, and while it has met with considerable success (specifically in the ancient world *viz* Rome), empire as a permanent bulwark against entropy has failed. Many times, when an empire begins to decline, it reverts back into the state that it was originally meant to defeat *viz* Rome 476 AD.

However, while Rome had fallen, the Eastern world under the Byzantine Empire began to thrive. However, the Byzantine "Empire" is a misleading term. While historians recognize it as lasting from roughly 330 AD until 1453 AD, it was by no means as influential as Rome. By 1204, it was basically the city-state of Constantinople. After the tremendous energy spent in trying to recapture Italy, the Byzantine Empire went on a permanent defensive and had to repel invasion after invasion from the Magyars, Muslims and even Christian crusaders from Europe, until it finally fell to the Turks in 1453. Once again, this illustrates how force held the empire together, and when the proper amount of force could not be maintained, the empire fell.

There is no universal formula for the failure of an empire. However, there are usually similarities, such as outside pressure and internal corruption. Empires are no longer a feasible government because they cannot assure the well being of all their citizens. [26] This however, is not to say that all citizens of all governments in the world today are taken care of. There are many totalitarian governments in the world today that do not uphold their citizen's human rights. I believe that modern totalitarian governments have reached an impasse. In the light of humanism, civil liberties and human rights, many governments still fall short. Instead they exist in a sort of stasis, akin to a mode of self-preservation, with a good chance of survival, but no real chance of advancement for the majority of citizens. An empire, and especially a totalitarian or authoritarian government reduces man to one subject, it strips man of his creativity and forces him simply to be an organism.

While I am reluctant to assign any special prerogative to Rome and the west, the fact that it lasted for other 1000 years does beg a closer examination. Rome was the pinnacle of an empire (although the Mongolians did control a larger territory, they only lasted a fraction of the time) because it controlled a vast amount of territory, and more importantly, it had longevity. And so, directly after the fall of Rome in 476 AD, while the Middle East was ruled by the Splendid Eastern Roman Empire, or the Byzantine Empire, and the Far East was ruled by various oriental dynasties, a change enveloped many of the former lands of the Roman Empire. Religion was paraded as the new bulwark against entropy. No one in the former lands inherited the ideals of empire because it failed. While the idea of empire was not dead, and various Roman ideas were incorporated into later generations, it never would achieve the success it knew during the classical era. Now, a new manifestation would be needed.

What was found was Christianity, which had enjoyed amazing success during the last throes of the Roman Empire. On a practical level, virtually all people adhered to Christian doctrines for a variety of overlapping reasons, the main one being the vacuum left by the absence of the Roman Empire. However, this vacuum had begun

---

[26] Some would probably make this claim about modern democracies as well, and it is a claim that I would unfortunately agree with.

to appear while the empire was still functioning. Many people failed to view the empire as a stable entity, and with the rapid spread of Christianity, they turned to religion.

However, while pagan religion had always played a crucial part in Roman society, Christian religion was vastly different. While Roman religion reinforced societal and ultimately earthy obedience, Christian religion dislodged it. Christian religion dislocated people from the earth; it forced their eyes upward, away from their own flesh. Now, the battle for existence had reached a completely new dimension. The battle was no longer an earthly battle, bent on preserving some measure of tangible progress, but rather, various intellects proposed that this existence be shunned. In its stead, they created a new existence, a spiritual paradise which was the embodiment of progress. A place in which chaos (symbolized as Satan) was completely annihilated. This new paradise could be achieved simply having faith in God. Earth was a stepping stone, as well as the human body, a place where people held back the inevitable chaos long enough to reach a divine state. The chaos that threatened them seemed too much to bear. There was no longer a stable government. Crime and barberry abounded in a once ordered world. While I am reluctant to make vast sweeping generalizations about time periods, this statement can be made with some certainty. Many former areas now existed in a vacuum created by the collapse of Roman political structure. People looked to the earth and saw only violence and self-preservation. However, on a conceptual level, what they saw was disorder; they saw entropy slowing coming closer like high tide. And they fought it back by doing the only thing they knew how to do, which was retreat higher on the rocks, to move closer to God. This situation roughly existed from 400-1000 AD; known as "The Dark Ages,"

This shift from earthly progress to spiritual progress not only enveloped the former Roman lands, but also many of the Middle Eastern lands with the emergence of Islam in the middle of the seventh century. Muhammad, a once wealthy merchant on the Arabian Peninsula, had a vision. In it he spoke to Allah through the angel Gabriel. Allah told him that his people were wicked and needed to be saved. Muhammad laid down his ethical laws, like an Arabian Moses, and shortly thereafter, many in the Middle East had converted to Muhammad's extreme ethical monotheism[27]. And so, by 700 AD, most of the European world and the Middle Eastern world did not look to achieving progress on this earth. To them, progress was not attainable in the flesh, rather, only in a divine realm. Whether it was Augustine or Muhammad, the souls of many people were forcibly torn from their bodies. Flesh was simply made to burn in hell. The earth plummeted into chaos while almost all prayed to God or Allah.

The antecedents to these paradigms lie mainly with the ancient Hebrews and ancient Greeks. The Hebrews, sometime around 2000 BC, began to strictly worship one god, Yahweh. Eventually, they disregarded all other gods in favor of Yahweh. They believed that God was not of this world, but rather had created it, in his complete omnipotence. This is a point in which I will revisit later. However for the discussion now, I will focus on the Hebraic vision of Yahweh. This vision was taken as the starting point for Christianity. A completely omnipotent and all powerful being created the universe and humanity as well. All power was thus centered in a non-human,

---

[27] Muhammad had originally been inspired by Judeo-Christian teachings which he had been introduced to on his business travels.

transcendent being. The Jews compromised the ability of humanity, and preached that all good must have been created elsewhere, in another realm of existence.

The Jews had prophesized that a savior would eventually rescue their race and elevate them. Around the time of Jesus of Nazareth, some Jews began to think that perhaps Jesus was their savior. Jesus was a man who uplifted the downtrodden. His homeland of Judea was a province of the Roman Empire. As has been mentioned already, under the Roman Empire, many people were alienated, driven further from the ruling classes and the power structure. Many of these disenfranchised people found hope in Jesus. Not quite a hundred years after Jesus' death, Paul of Tarsus (later St. Paul) began to preach a radical doctrine, which Jesus was not just a savoir, but that he was God himself, he was the universe and the eternal logos nailed upon the cross. Many Jews, even ones that accepted Jesus as a savoir, could not accept that he was God. This point of contention was a major factor in formation of Christianity. And so, the "Christians," the ones who believed that Jesus was the savoir and God himself were born of the Hebraic faith.

From the Judaic tradition, coupled with the life and teachings of Jesus of Nazareth, the west inherited a synthesized omnipotence, will and order. They then applied these concepts to government and society in an effort to combat the ever encroaching disorder, as well as to a fill a void left by the collapse of the Roman Empire. Now, progress was guaranteed in the next life. And it is easy to understand how this idea was manifested and believed by almost all. Upon inspection, the earthly realm was a cesspool of violence. Men and women had to struggle in a violent world dominated by the rich, which they were left to starve. The European population living from 400s-1000 AD came face to face with disorder, both in reality and concept. Indeed, these people could not have believed that progress was capable on this earth. And so, the notion of transcendence, of Jesus being the omnipotent, all powerful creator the universe was manifested and believed by almost all segments of society. Progress was attainable, just not in this realm. A paradigm of order was extrapolated from the bible and Christ teachings. Order and morality were pressed up the people through guilt and fear. The fear of hell, the fear of disobedience loomed for many medieval men and women.[28] People now had a model of order to abide by.

However, with the impressments of morality and conduct on the people, they were also impressed with divinity. A power greater than themselves, perfect in every respect. This was not a power they had within themselves however, in fact it was something their material bodies were incapable of achieving. This divinity was not the old power of the scattered gods of Egypt, Persia, Greece or Rome. All of their power was rolled into one god. Yet, this power was tamed, easily accepted. Order was a rosary away. No one bothered to consider what would happen if this power of divinity were not tamed. I will visit this point later on.

As mentioned before there was a similar occurrence in the Middle East. From the backwardness of the Arabian Peninsula, from its being sandwiched between Roman influence and Byzantine influence, Islam propelled it as a powerful entity. Both Islam and Christianity became governmental structures; both preached a transcendent method for the attainment of progress. I will discuss the progress of Christianity in the west and then the progress of Islam in the Middle East.

---

[28] Not that all were in fear, see Cantor.

The "Dark Ages" is a controversial label given to most of Europe from the fall of the Roman Empire in the fifth century after Christ until the beginning of the Crusades in the end of the eleventh century. While it is hard to pinpoint a specific label for a region and time, what can be said is that people in this time period did not enjoy a very high standard of living. There was virtually no education, medicine or governmental stability. All the while, religion maintained any semblance of coherence in society. My aim is not to discuss the evolution of Christianity in the Middle ages, but there are nonetheless a few crucial points to my discussion. For one, in 590 AD, Pope Gregory the Great assumed responsibility for collecting taxes maintaining some order throughout the city of Rome. This was a crucial event, for it illustrated the new form of government, which in essence was a theocracy. The feudal system which gave some order to medieval society was completely endorsed by the church as well. Perhaps one of the most important functions of the church was its preservation classical culture. As the barbarians and Vikings terrorized Europe, education came to a standstill. The vast amount of knowledge of the classic world was preserved in church libraries and monasteries. During the beginning of the middle ages, feudalism, political popes and clergy and church scholars all served as the bulwark against entropy. It was in these institutions where some semblance of order was preserved, even if it was only for this earthly realm. Religion assumed the role of guaranteeing order in a chaotic world. Not only a governmental level, but on a social and individual level, religion became a source of community and hope amongst desperate people.

However, as we know, religion as bulwark against entropy failed, both in the European world, as well as in the Muslim world.[29] Around the year 1100 AD, a claim can be made with some certainty that the Catholic Church was at its peak of power, at least in Western and Central Europe. By 1100 AD, the first crusade was under way, a struggle in which united much of the population in the west against the Muslim world. Hillenbrand, later pope Gregory, established the supremacy of the Catholic Church in secular and political affairs, as well as religion. The church's sphere of influence extended into every aspect of society. For generations, the population of Europe "inherited" the doctrines of the church. Even if not everyone believed in those values, the values nonetheless permeated the fabric of society. The church lead the battle against entropy, it gave order and meaning to all parts of life.

Yet, after 1200 AD, the church began to weaken for a variety of reasons. The failure to capture Jerusalem and the subsequent defeat at the hands of the "heathen" Muslims, the growing corruption and flagrant abuses of clergy, the horrors of the inquisition, the rise of education a higher standard of living. On a practical level, this meant that the church lost the monopoly of influence that it held in all aspects of medieval life. Politically, secular governments (namely monarchs) began to assume political control where the church formerly had. After the concerted war effort needed to maintain the crusades, as well as the one hundred years war, secular monarchs assumed control because of the stability they offered to warring nations. Socially and culturally, with the rise of universities, and other schools, more people were equipped with the ability to reason and logic, thus, they were able to question the dogma of the church which had been untouched by the illiterate populations that had preceded. The emergence of the Black Death in the 1340s was a crippling blow to the prestige of the

---

[29] What about religion as bulwark in the Far East?

church. Obviously it was sent to punish humanity, but prayer and homage became useless, so did the church itself, nothing stopped the onslaught of the plague. Priests abandoned their parishioners to the save themselves. All in all, the plague killed almost 40% of Europe's population. The church was meant to be an edifice of order in a chaotic world and in a time when it was needed it most, it did not provide any answers, and it did not provide an order.

However, a major factor in the weakening of the church was economics. Financially, from the trade sparked by the crusades, more people had more money, thus leading to a higher standard of living. Historians usually term this as the rise of towns. Norman Cantor, in his "Civilization of the Middle Ages," states that people who now had money did not want to believe that life was the rotting cesspool of violence that they had been taught. I agree with this assessment. As people became learned and wealthier, I believe they came into conflict with the inheritance that they received from their ancestors. They did not want to completely forsake this life, or tear their souls from the body. I believe that with the weakening of the Christian church, and the general rise in the standard of living, people wanted to justify their earthly life. And so a new manifestation was needed.

The dissatisfaction with the Catholic Church found its patron saint in Luther. In the early 1500's Martin Luther had shattered the Catholic Church by creating his own church, Protestantism. What was a point of contention for Luther was not so much the corruption of the church, but that very foundations that the church was built on. For one, the institution of the church and the clergy were supposed to act as a mediator between the laity and the divine. Luther wanted all people to have a direct access to the divine. For our discussion, we can view Luther's desires as him wanting every individual to have access to that sublime progress which was the reality of the divine. He did not believe that certain individuals had access and were responsible for "relaying" the divine to everyone else[30]. Luther shattered the unity of the church and fatally compromised its ability to be a bulwark against the entropy.

However, too many times history is presented as a neat, clear cut entity with clearly marked divisions. I do not want to fall into this trap. The weakening of the Christian Church from the 1200s to the 1500s did not abolish religion. Even in the present day, religion is a means for many to combat the disorder of their lives. As far as my economic analysis, I do not believe that everyone's standard of living dramatically increased. Poverty still abounded, and I would venture that a vast majority of the people in this time period were devout Christians (either Catholic or Protestant). But what was obvious was that the Catholic Church had lost its hold on society. After the sixteenth century it could no longer permeate every aspect of society. Its ability to do so had been compromised by the above mentioned factors. The values and ideas of the church, even if now not completely trusted or adhered to, would be inherited by a new generation. However, this new generation would need to seek another defense against entropy. Empires and Religion had failed. They had

---

[30] Although, there is a major contradiction in this, because while he did not believe in the power of the Catholic Church and clergy, and while he wanted a direct relationship to the divine for individuals, he also preached a doctrine similar to Augustine, one of predestination, or grace. However, it is not my aim to dispute Luther's views. I only aim to highlight this contradiction.

served the populations well, and had been successful for a while, and would continue to be influential in limited ways, but they had been exposed. Their contradictions and shortcomings had been paraded around for all to see.[31] Despite weakening, the church would leave its inheritance for the later generations. Namely after its decline, the Christian church imprinted western society with a synthesized omnipotence, will, and order. The west inherited an attainable infinity, a tamed transcendent order, not truly known to the pagan traditions.[32] Yet, when this transcendent order was applied to society, it became a rigid, dogmatic system, a far cry from its embedded transcendence.

I will depart momentarily from Europe to visit the Middle East. From 622 AD until roughly the middle of the 1300s, the Muslim religion had served a similar purpose to Christianity in the west. While they spent centuries fighting each in bloody battles, I believe that what they each could not recognize in the other was their passion for order (despite being a static dogmatic order). In fact, by battling each other, they actually descended further into what they were trying to prevent. After the death of Muhammad, Islam spread rapidly throughout the majority the Middle East, through parts of northern Africa and even into Europe, as far as France. Islam stabilized the Middle East and was responsible for the growth of a rich culture, of mathematics, of astronomy, of medicine, and of reinterpreted western philosophy. Yet by the 1300's, a fundamentalist movement had taken root, possibly due to the outside influence of the Crusades, possibly due to the growing secularism of the Uuymad and then the Abbasid dynasty. As the west began to grow financially, culturally and militarily after 1300, Islam "turned in" on itself.

I do not wish to demean Islam, but from a western perspective, the religion of Islam seems to be one of static progress. However, I can not judge the Middle Eastern culture from a western perspective[33]. What seems like fanaticism and dogmatism to the west may be one culture's defense against entropy. The importance of oil and the billion dollar industry that has spawned definitely has had a major impact on Islamic and western relations. This is a point that I am not prepared to discuss in my present paper. However, I do speculate that religion-in the perspective of many Muslims living in the Middle East-has not failed as a bulwark against the spread of entropy. And it is here that my cultural ethnocentrism reveals itself. From my studies and personal experience, a religious government can at best achieve stasis, but not progress. Dogmatism regulates people to a subservient position of obedience. Obedient people have an extremely difficult time producing last achievements to combat chaos and eventually overcome it (not to say that it is impossible). When the means of education are controlled, it is easier to keep people subservient. Alex De Waal puts forth this theory. He states that education in many Middle Eastern countries is extremely scientific and rigid. Critical thinking, analysis and the humanities are almost entirely left out. This could be a possible reason for the differing of the western and mid-eastern cultures.

---

[31] However, not all believed or accepted the shortcomings of religion, even to this day.

[32] This is not to say that the pagan traditions did have a clear conception of the divine. Rather, I speculate that with Christianity, this vision was synthesized or combined into one, omnipotent all powerful deity, and thus more easily communicable to society.

[33] This is from Herders argument of cultural relativity and ethnocentrism.

In the Far East (India, China and Japan), the two main religions that took hold, even earlier than Christianity were Hinduism and Buddhism. Hinduism developed around two thousand years before Christ. Basically it is a religion of infinite chances. The main objective is to reach motska, or a union with the divine creator of the universe. According to one's Karma, or actions in life, ones position in his next incarnation is determined. This is the process of Samsara, or reincarnation. If you have good Karma, you go up a level, bad karma, down a level. This religion is one of the oldest in the world. Buddhism is a derivative of Hinduism, started by Siddartha Guatama. It states that the ultimate goal is a compete cessation of both being and non-being, this is a state of Nirvana (as opposed to Samsara).

Hinduism offers a static bulwark against entropy, but not a progressive one. By positing a hierarchal system (a caste system), dependant on Karma, Hinduism gives a certain measure of control to the participant and gives him a clear conception of order, in the union with Brahma. However, it demeans its participants by rigidly stratifying them and of by no tangible cause of their own doing (because if you are an untouchable, or of the lowest caste, the belief is that you must had done something in you previous life of which you have no knowledge of). The lower classes, which are the more numerous, suffer terrible indignities at the hands of the upper class. Hinduism promotes a sever classism which paralyzes society. While theoretically it is supposed to lead all to motshka, in reality, it only leads the rich and wealthy upper classes that benefit at the expense of the lower classes.

While Siddhartha Guatama opposed the rigid distinctions extreme prejudices of the caste system, Buddhism actually is a religion of submission to chaos, and ultimately nothing. Nirvana is *nothing,* it is not a state of being or non-being. When reached this perfect state of understanding, there will be no more divisions and discriminations like in the earthly realm. There will be infinite understanding, but this infinite understanding is equated with the cessation of all being. True Nirvana cannot be described, but since it is nihilistic in nature, it actually is a submission to chaos, a relinquishment of order, a capitulation of reason and progress[34]. Yet, the contributions of Buddhism and Hinduism are immeasurable. Buddhism brought man face to face with the void; much like Luther did 2,000 years later. Siddartha saw the nothingness; he felt the ever pervasive threat of disorder.

So after the initial weakening of the Christian church in Europe, the question is what could replace it? During the 1500s, traditional European culture was rapidly changing. The power of the church still had a considerable hold over the minds of the populace, but it was rapidly weakening, especially in the eyes on the elite and the intelligentsia. Due to the influx of wealth in the Italian city-states after the crusades due to their location in the trading network, the rise in educated people, the subsequent abuses by the church, and the effects of the reformation, people began to view their world differently, as well as themselves. Many began to appreciate this earthly realm (which for so long had been admonished). Instead of merely surviving and hoping to achieve order in some ethereal realm, this new movement, later dubbed humanism, began to place emphasis on the secular world. Whether this was a justification for the opulent lifestyle of the new merchant class, or whether this truly was a genuine shift in

[34] Pope John Paul II

opinion (and I believe it was both), people began to realize the value of human life, they began to value the individual, while still trying to achieve a divine progress. What is significant of the humanistic movement is that the attainment of order was extended to include the earthly realm once again.

While religion was being exposed of it's shortcomings (at least in the west), it nonetheless imprinted its values on the next generation. What the succeeding generation inherited from their Christian heritage was the idea of divinity and of the Judeo-Christian conception of ethics. Divinity for our purposes will be defined as the realization of all potentialities and possibilities, located in an infinite existence. However this existence was not accessible to a majority of the population. People subscribed to the church and the clergy for their fix of order in a chaotic world. But the weakening of the church coupled with the new growing humanistic currents made people crave order, but not a watered down dogmatic order, but rather an order they could build themselves. Divinity started to become a human thing. This can be evidenced in the spectacular works of the Renaissance, Leonardo Da Vinci, Michelangelo, and Machiavelli, all began to define formerly religious scenes and occupations in humanistic terms. Government, the creation of man, all had been grounded in the human. The fight for order in this chaotic existence was now taken up by man. Although this judgment may be premature, man-even if a minority-had realized his worth, had realized his capability in the battle. He craved divinity and did not want to wait until they died.

This statement will most likely be disputed. I do not intend to lead a long-winded historical discussion. What can be said with certainty is that from 1500 until 1800 there was definite shift in the worldview held by a sizeable minority of academics, philosophers, artists and other intellectuals. And the logical outgrowth of humanism was total faith in reason, a total faith in the human ability to decipher the universe. The Enlightenment saw reason at its peak. Humanism put man equal to god. But once man envisioned equality, he then resented this. He wanted superiority. Man had has his taste of the divine. He painted works of art, wrote treatises, he sailed the ocean, he learned the laws of his universe. Why couldn't man be divine? Why couldn't man take order into his own hands, make his own fight. And so the unbounded faith in reason flourished. One Enlightenment thinker once confidently stated that in time, men would know the position of every particle in the universe. Men of the age of reason wanted freedom from superstition, wanted humanism, they wanted education and above all, they wanted order. Once again, I am generalizing. However, the goals I listed above were common to many men of the Enlightenment; these goals predominated many discussions in academic circles and were sought after by numerous intellectuals. While not all enlightenment philosophers agreed[35], many did agree in the power of reason to uplift humanity. They had inherited the failures of empire, and religious dogma. Many intellectuals wanted to build a world with reason.

---

[35] At this point I have to mention the Marquis de Sade. While many may balk at my inclusion of such a perverse literary figure, I admire the basic implication of his work. Whether he consciously wanted to raise this issue, his work begs the question: "what happens if we let freedom go too far?" Yet it also sends a warning of slavishly following dogmatic fanaticism. I hope all can forgive the tired cliché, but de Sade put humanity between a rock and a hard place.

In this respect, order needs to be seen as on the verge of victory. Some men of the Enlightenment claimed an early victory. Order, the ultimate underdog, which had been expunged from chaos, which had been taunted and almost vanquished, now was ready to deal a deathblow to its hated enemy. Yet, at the moment of its ultimate triumph, at the moment when it appeared that human reason would erase chaos and build its perfect world, something happened. Chaos, the resilient, fought its way back from under the foot of order. Men like de Sade began to punch holes in orders seemingly impregnable façade. Another man, David Hume, along with other "mercenaries" Hamann and Jacobi, posed fatal questions. How human beings can be so sure that our laws actually work. We assume that they will work, we build our civilizations like they will work, but how do we know? How do we know the sun will rise tomorrow? How can be sure that stove is hot when we touch it? The only thing the mind knows is its own experiences, which it then assumes are infallible laws, but which are only probably at best. We assume connections between cause and effects. In essence, humans assume, especially after the deciphering of supposed natural laws (in light of Newtonian physics) a certain degree of order that is inherent in nature, but really, this is an illusion, reason is an illusion. Hume, Hamann and Jacobi also believed that for anything in nature to actually function, what occurs is a type of faith. Humans invest a certain amount of faith-which they called Glaube-in functions and occurrences, and it is this faith which enables reality to function.

David Hume threw the foot of order off of chaos's neck. Like a mercenary, he saw a weakness and exploited it. Reason staggered like a stunned boxer who had been knocked down by someone he almost beat. The fight was not over. New questions then arose in light of reasons' now exposed weaknesses. Was reason an objective standard of truth for everyone, or was it subjective and conditioned by culture and experience? Basically, is a human born with reason, or is it a result of his experiences, which are subject to contamination and vary from person to person? Reason began to falter. Chaos laughed at its cocky opponent. If one followed the path of reason, where would it lead? Many posed it would lead to nothing. How could reason explain the creation of the universe? If we assume that everything has cause, how could the earth have been "reasonably" created? That would impose causes ad infinitum, or an uncaused caused, both of which are irrational. Was reason unreasonable? Logic illogical? Unless we "took a leap of faith" and blindly believed in a higher power we would be left with nothing. The inheritance from religion, which was on the verge of failing in academia, now beckoned. But order refused to lapse into dogmatism. Nihilism waited for order in one direction, dogmatism in another. As Frederick Beiser has so named it, what would be the fate of reason?

At this moment I would like to bring our discussion to the masses. What did any of this mean to them? Did these abstract philosophies feed them? Give them shelter? Protect their children? Obviously not. What about the Muslim world? And the Far East? What about a poor farmer tilling his fields six days a week for a feudal master while feudalism was supposed to be dying? My point is that this battle over reason did not affect a greater part of the world. However, I believe indirectly, all are affected by it. The waiter who prepares your coffee in the morning, the doctor who performs open-heart surgery, even a terrorist, they all fight disorder; they are all part of the battle. A Buddhist monk trying to reach Nirvana is striving against disorder in his own way.

While I focus on the intellectual battle, I want to make it known that the battle took almost infinite forms. For millions of years generations of humans had come and gone, they had evolved, all (supposedly or theoretically) better than their ancestors. Humans hunted, killed animals, killed each other, and traded seashells. Order so desperately tried to pass its secrets and its weapons to later generations who could wage a better battle, perhaps vindicating order at some later time in the future. All humans struggle against disorder, but this struggle was articulated in various academic circles.

So, what was the fate of reason? How would order respond to the shattering blow dealt to it? Philosophers after the Enlightenment debated this point relentlessly. While cities boomed, while Christians fought Muslims, while people toiled in the fields, while the Ottoman's clung to empire to stave back entropy, two post-Enlightenment philosophers arose in Germany. Hegel took his enlightenment inheritance and prophesized a world of rational progress, a teleos which would culminate in an absolute state. The goal of history was the evolution of the absolute spirit, it was a drive toward freedom. How could the world not be moving forward? He pointed to the progress of human history. Look at the evolution of governments and technology and philosophy. Misery and disorder were only parts of a greater unseen pattern, one only known by god. Schopenhauer, the contrarian, took the inheritance of the enlightenment and posited a world or irrationality, a non progressive chaotic aberration, reminiscent of the feared nihilism of his earlier contemporaries. How could an irrational world not be so? Schopenhauer asked this question with his works, which made reason subject to the will. Reason was not free, it is a sickly puppet of the irrational, bent and twisted to do the irrationals bidding. So who was correct? The battle pitched. Was reason a subjective truth, a malleable thing determined and manipulated by a blind irrational will? Or was reason a divinely posited thing, moving humanity ever forward?

This question, I believe, is still being played out. As general bits of Hegel, Schopenhauer's and their contemporaries (i.e. Darwin and Marx) theories tricked down to the masses, the masses inherited a conflicting battle. This conflict was then grafted onto the failures and contradictions of earlier inheritances. Reason was a casualty in the sense that it became a means to order, not an end. And despite the limits of reason being exposed, many chose to forget this and exalt reason.

During the late 18th and early 19th century, a small poetic movement arose in reaction against the Enlightenment at its values, primarily its exaltation of reason. This new movement, called Romanticism, aimed to counter the cold, mechanistic prison cell that reason placed man in. Many also opposed the harsh conditions of the new industrialization. What began with Hamann and Jacobi soon became a full fledged movement. But unlike Hamann and Jacobi, Romantics did not simply use religion to rebut Enlightenment, instead, similar to their Enlightenment opponents, they also sought to rebel against imposed authoritarian structures. Yet, instead of relying on reason, they relied on imagination. And in an ironic twist, many Romantics were influenced by the theories of one of the greatest enlightenment; Immanuel Kant (who wrote prior to Hegel and Schopenhauer). Kant proposed that the human mind is an active entity, that it actively structures reality into a knowable, familiar and perceivable entity. The Romantics expounded on the idea of this creative and imaginative structuring of the universe. I would like to pick up on this trend. I believe the implication of Kant's discovery is paradigm shattering. Simply put, Kant deduced that human beings *contribute* aspects of reality, thus making it their reality, making it subjective. Humans are born with certain cognitive attributes which then structure reality accordingly. Humans contribute, add to, infuse and impregnate reality with their own faculties. I will leave this point for now and discuss it in the conclusion.

And it is here where the Romantic Movement falls neatly into this discussion. I want to extrapolate this idea. The Romantics believed that existence was not a thing to be conquered, but rather a mystical entity which only imagination could penetrate. Nature, and existence itself was infinite, and cold hard reason fell stillborn in the face of existence[36]. However, Romanticism is crucial for our discussion because while it never truly influenced a major portion of society, it nonetheless put man in touch with the key to his own liberation, his own happiness, his key to producing the order which every cell in his body is geared toward. But it is not a false or temporary order, not static illusionary progress, but rather, one saturated imagination, one that will fulfill him infinitely.

Obviously, society did not crumble, reason, despite being exposed did not vanish. The exposing of reason was only in elitist philosophical circles. It had no true bearing on the daily lives of the masses. Reason was slowly infiltrating society and replacing superstition. Hand in hand with reason, went science. The Scientific revolution, as it had been termed, began in the 1400s with Copernicus. Copernicus had postulated that the earth revolved around the sun. This was no doubt an outgrowth of the "mini-Renaissance" of the 12th century during the high middle ages. Kepler and then Galileo eventually took up Copernicus's cause and eventually proved the heliocentric theory, thus dislodging man from his privileged position in the heavens, ordained by god. Science, reason and criticism were all married together. This new holy trinity would break the old reins of superstition and dogma.

Indeed, we still live with this inheritance. Since the 1400's, science has transformed society. After Galileo proved the heliocentric theory, the door was opened and other thinkers began to systematically analyze the natural world. Enlightenment reason was applied to the world and its natural laws were sought after. Formerly, the laws of the universe were simply attributed to God. But now, human beings were increasingly realizing the ability within themselves to decipher the laws of nature. Sir Isaac Newton's theories revealed nature's basic laws. God was regulated to the sidelines, he was the "eternal clockmaker," who set the universe in motion and watched his creation. He was not longer in involved in his creation. In the 19th century, Darwin proposed his theory of evolution, which then dislodged man from his position as divine species, now man was an animal. Freud exposed man's irrational subconscious instincts (much like Schopenhauer) and Einstein helped to develop one of the most powerful forces ever unleashed by man.

While I am only highlighting the major scientists and their achievements in the last 500 years, there were many others. The scientific achievements of the last 500 years rigorously examined reality, held it up to the eye. It seemed as if the intelligentsia were no longer content with merely attributing all glories to God, man

---

[36] Later philosophers, especially existentialists, explored this idea; they probed the depths of existence, and some found nothing. Some found only dead rules but no true existence. They claimed we must make our own existence. I will expound on this idea.

wanted a share for himself. The universe became wide-open and accessible for the human mind. While many people still adhered to religion, others now used a new means of deciphering their world, of illustrating the battle between the entropy and order. In our present society, the belief and the reliance in science is manifested by the majority and accepted by most. Science has transformed daily life for almost everyone on this planet in 2007. Science gave man a new weapon to expose and rout his enemy of chaos. In the eyes of some (not all) science could almost be viewed as new savior of mankind. Various attempts were made to keep it separate from religion, but the rationality of science eventually overtook the revealed and dogmatic position of faith. And yet, the same dilemma is with us, waiting to be solved. If reason is traced back to its source, it becomes irrational. If we believe the entire universe to be governed by immutable laws of cause and effect,[37] then from where did the first cause come from? Various thinkers during the Enlightenment tried to defend the objectivity and superiority of reason, but they failed. Reason could not stand up to this mighty argument. Thinkers looked back to metaphysics, nature and transcendence to explain reality where reason had failed.[38] And even today, we either have to embrace skepticism and doubt everything, or take a leap of faith and believe in an uncaused caused. Science, for all its achievements, still cannot answer the age old questions of why we are here, or what happens when you die. While many would deny the infallibility of science, no one can reasonably postulate the answers to these questions. Perhaps, in time, science will be able to bridge the gap between faith and reason, and answer some of these age old questions. But as of the present, these answers still escape us.

Another criticism of science is one that had faded into the background, but is no less fatal. Albert Einstein developed one of the most powerful theories known to man: nuclear power. When the atomic bomb was unleashed on Japan in 1945, the world changed in moments. Hundreds of the thousands of people died in an instant. While this was horrific, the metaphysical significance of the bomb was even more drastic. I stated earlier that vast new avenues of exploration had been opened up within reality and the natural world. Men were optimistic at what they could discover. But after August 6th, 1945, only a sadist could be optimistic about what was found. Humanity now had the capability to completely annihilate itself within moments. Science, the savoir of order, the guarantor of progress, now could obliterate any sign of human life. Science now played both sides of the fence, chaos and order, along with reason. Reason and science were no longer ends; they no longer could vanquish chaos. Instead they became means to an end, means that could be hijacked and abused by either side.

However, the atomic bomb was only the culmination of this failure. As early as the sixteenth century, many began to resent science for its mechanistic and reductionism tendencies. Science forced man to become simply a piece of nature, stripping him of creativity, emotion and self expression. Many reacted against this mechanistic concept; they tried to preserve their humanity in the face of science and later industry. Various movements, most notably the romantic movement of the 19th century emerged, trying to reaffirm man's humanity, preeminence and creativity.

---

[37] Also see Hume's argument of causality
[38] This was a major contributory factor of the Romantic movement in Germany

Once again, I must clarify my argument. I do not propose a linear progression of scientific achievements, and an inverse regression of religious adherence. All I propose and I believe that many would agree is that in the last 500 years, scientific achievements have definitely triumphed over religious achievements. Religion, in the west, shattered under the weight of its failed promises, its dogmatic claims, its internal divisions and its massive and widespread corruption. Many have turned to science. Its promise of logic and reason are comforting to an increasingly more educated society. Science has infiltrated daily life. Religion still plays a major role for many, but no longer in a political or communal sense (at least in the west). Religion has been regulated to an individual, private matter separated from the state in many areas of the world[39]. Cell phones, internet, dishwashers and automobiles are norms in modern society. And I believe our modern technology can be put to uses that many never dreamed of. Technology must be applied and used correctly, it can aid us in the struggle. Our society is built on the pillars of science and logic. Yet these things are not ends. They are means. Order and chaos still circle each other; they still square off, face each other, pant and spit blood. One is trying to vanquish the other; one is trying to rape the other into submission.

European historians sometimes refer to the 19th and early 20th century as the age of nationalism because of the formation of various independent nations. However, I think the idea of nationalism or allegiance to a particular group is an idea as old as the intellect itself. However it is not my intention to debate the age of nationalistic tendencies. Instead, nationalism is crucial to this present discussion. Nationalism is a key defense mechanism in fighting off entropy. A group of people align themselves together, based on race, language, history customs or any uniting factor. Various people adhere to this "uniting factor" and it can become the basis of a foundation.

In the 19th century, various new nations formed. Some formed on the basis of ancient prerogatives, others on modern ones. Some nations formed from geographic or cultural ties. Still, other regions had been incorporated in various empires had fought for their independence. Greece, Albania, Italy and Germany were among the new European nations that formed, along with the various nationalities of the splintering Austrian empire. Later in the 20th century, Poland and Yugoslavia formed. Populations in these areas could now adhere to a new bulwark against entropy. This was a social and communal institution, a unifying force in the lives all its adherents.[40]

"Italy" had been a collection of independent city-states, papal lands and foreign controlled states since the middle ages. Eventually, the kingdom of Sardinia-Piedmont under Count Cavour began to unify many of the smaller lands in between.[41] With the rising swell of nationalism and autonomy due to the French revolution and the Enlightenment, various liberation organs formed within Italy. The most notable became the liberation army of Giuseppe Garibaldi. With Cavour's diplomacy and

---

[39] This obviously does not include many parts of the Middle East, which I have addressed earlier.

[40] this is not to say every member of a new nation embraced nationalism, some were hurt by it or remained unchanged

[41] Whether this was for Italian unity or Sardinia-Piedmont hegemony is a contested issue.

armies pushing from the north, and Garibalidi's "black shirts" driving from the south, the country was eventually unified by 1871.

Germany, under Prime Minister Otto Van Bismarck followed a similar pattern at roughly the same time. "Germany's" largest state was Prussia-Brandenburg. Second was Bavaria in the south. In between were over 300 German principalities, which were reduced in number to 39 after the Congress of Vienna. The French Revolutions of 1789 and 1830 greatly contributed to the nationalism of the major German states. After a failed attempt at a constitution for "Germany" (which was not unified yet), Otto von Bismarck became prime minister. His program of "realpolitik" (no idealistic politics), industrialization and militarization made Germany a formidable power. Bismarck led three wars, one against Denmark, one against Austria and one against France, all for gaining territory. By 1871, Germany was a unified country, and one of the most powerful in Europe. Germany's dominance would be one of the main contributing factors to World War One. Their vindication of their defeat in world war one helped to spark World War Two. I have chosen to give the details of Italy and Germany because their formations were crucial to European history and to world history. At each "step," various people in the respective societies demanded their countries privilege to be a country. This privilege then became a right, and it became a right at others countries expenses.

Italy and Germany's path to nationhood are of utmost importance to our discussion. Even though I omitted many details, their paths have a greater significance. What is nationalism? It is the feeling of belonging to a particular group. It is pride in ones heritage. It can be an amazing bulwark simply because of its uniting factor, for its ability to persuade people to work toward a common goal. During their drive toward to nationhood, many "Germans" believed in their shared heritage, as did many "Italians". They rallied around a common element. And it is precisely at this point that nationalism is also one of the biggest detriments to order. When one nation is convinced of their superiority, of their right to inhabit a specific land, when one nation is convinced of its right to uphold their vision of order, they band together and wage war against another nation or group. Nationalism is not a new idea by any stretch of the imagination. The Greeks, Persians, Romans, French, English, Italians, Germans, Americans, Yugoslavians, Russians, Chinese and Indians all have had a common rallying point of culture and heritage. And at various points in all the aforementioned nations' histories, whether it is antiquity or yesterday, a war has been fought in their name. A battle has been waged to uphold the natural right of Germans to take land or of Americans to control oil fields. Nationalism can exalt one group and at the same time, destroy another. This is order and progress for some, according to a perverted vision. But this brings us to the question: who has the right vision? Who is to say the Germans were wrong?

Undoubtedly, my last point will horrify people. However, the discussion of nationalism has forced me to deal with a larger issue. If progress and order are sought by many, inevitably, conflicts over the nature of this progress or order will arise. In the case of Germany during World War Two, of the hijackers on September 11th, a discrepancy arose over their vision of order and progress with that of most of the rest of the world. So, how do we determine who is "right?"

First off, I think our word usage must be clarified. "Right" is a slippery term. Indeed, Hitler thought he was "right" in doing what he did. A subjective basis of "right" or "good" could lead to disaster (as it has in many cases). I believe that an objective standard must be employed. However, I do not propose a standard of right or good, but rather, an objective standard of order. I believe that this standard is necessary in order to prevent certain individuals and groups from forcing their versions of order on weaker individuals or groups[42]. The way to achieve this standard would be by simplifying it into a type of formula. It is utilitarian in nature, but I would not succumb to Mill's "tyranny of the majority." A progressive standard of order must be employed to ensure that as many people as possible are led to order. I believe that this objective standard of order is evident in the simple fact of birth or creation. For what other purpose are organisms brought to life? The answer is to survive, to rebut entropy. Indeed, the functions of the sperm and egg, than later the zygote and infant are all bastions of order, albeit an unconscious order, but order. Despite the corrupting influences a person may succumb to during the span of their life, they were all born, and it is their birth which is the testament to the objective, natural standard of order, on inherent in all living creatures.

The individuals that do not adhere to order voluntarily succumb to chaos. As mentioned earlier one, not everyone will choose to follow order. Thus, the formula could be read as $ax=-bq$. As the amount of people striving for order increases (the people are represented by x, their increase represented by a,) the amount of people striving for chaos will hopefully decrease (people who adhere to chaos are represented by q, and their decrease it represented by –b). I do not venture that this formula is rigid or applicable in all cases. Many will no doubt find fault with it. However, all I present is the attempt to solidify order and progress for as many people that truly understand what it means to be progressive or have order. However, I do not believe society is ready to achieve a higher state of progress yet. Institutions must be developed to teach society to achieve this state of order. I will expand this point at a later part in my discussion.

Nationalism has failed to lead humanity to order. Especially after World War One and World War Two, the destructive capabilities of nationalism bore fruit. Nationalism failed to become an end, but rather a precarious means to that end. A means which many thinkers have disagreed with (i.e. Marx). The last idea I will examine is one that has irreversibly changed society, it is one that affects all people, whether directly involved or not. It is an idea that has transcended its original boundaries, and has been applied to many institutions in society, for better or for worse. I have argued the failures of various defenses against entropy, starting with

---

[42] Some may object to my treatment of "the weak," or ones who are unable to sustain themselves in the face of dominance. They may sight the rules of nature do not permit such safeguards. However, I retort with the fact that the person writing those words most likely is doing so from a relatively comfortably setting, plush with societies conveniences, protected by police and the military. Many who adhere to Darwin's laws are hypocrites because at the first instance of being attacked, or perceiving and attack and a loss of their possessions, they look to the police, or if they perceive a threat to their land or economic interests, they call loudly for the military. If Social Darwinism is followed to its logical extreme, it would result in the breakdown of society. Order would disintegrate. Indeed it is one of the hallmarks of our intellect, of our very humanity, to care for the weak and to love. While this notion existed before Christianity, I believe that Christianity imprinted this idea onto many later generations in the west, and subsequently has become part of our inheritance in the west.

governments, religion, reason, science and nationalism. However, while I do not contend that this next method had achieved a victory, nor do I believe that it can achieve a victory, but rather, out of the previously mentioned methods, the one I am about to discuss has had the most success of unifying people. However, more times than not, humanities unification by this process has been a negative one, or in opposition to it.

During the 1790's Great Britain underwent what later became known as the Industrial Revolution. New methods of commerce and production emerged as a result of this industrial revolution. Various factors contributed to the revolution, including but not limited to, the availability of raw materials, the fencing in of private land, the small size of the Britain, the demand for goods, laws protecting industry, the involvement of government and the military in the sphere of industry and commerce (mercantilism), as well as the emergence of new technology. Britain fast became the world leader in industry. Not surprisingly, Britain also began to colonize lands across the globe. From the late 1790s well into the 1830s, Britain remained the world leader of industry. Armies of urban workers filled the newly built factories, coal shipments steamed in, the face of British society transformed. Wealth accumulated and remained frozen in the hands of landlords and factory owners, or the new bourgeoisie or middle class[43]. However, while the noble class and the middle class rose in wealth and status, they left the new working urban class behind. The "proletariats" as they came to be called crowded into little hovels with barely enough substance. They had very little access to education, so social advancement was almost impossible. While I do not intend to paint a black and white scene of this newly emerging struggle, and while many people were unaffected at the time, industrialization had a permanent and irreversible effect on English society. Later, France, Germany and the United States industrialized as well.

What happened during industrialization? For one thing a gradual switch from handmade goods to manufactured goods emerged. Countries military capabilities dramatically increased as well. But most of all, and the most important for our discussion, is who benefited from industrialization. As I mentioned, a new bourgeois class emerged. This class mainly consisted of factory owners who controlled the means of production, as well as bankers, merchants, bureaucrats, clerks and lawyers. The bourgeoisie in England became immensely wealthy on the backs of wage labors, whose wages could barely sustain their families. The petty bourgeoisie also emerged, such as shopkeepers and smaller merchants.

While it is not my aim to discuss the emergence of industry, and while I have glossed over some important points, industrialization can be reduced to a few notable points. What can be said with certainty is after 1790 and into the 1800s, new social classes developed in England due to the effects of industrialization. The aristocracy or landed class (which was not new), the bourgeois, petty bourgeois, and the urban workers, skilled and then unskilled[44].

---

[43] Landlords were usually of the old nobility class. Vertical movement from the middle to the noble class was rare but it did happen. Vertical movement within the middle class was common.

[44] There were also still a sizeable portion of the population that worked from home (cottage industry) and that worked as agricultural workers.

Gradually, societies began to value quality over quantity. Industry had imprinted itself onto the very fabric of society. It gradually transformed society by regimenting it. Strict adherence to time became crucial to the operation of factories. Rigid hours were kept by laborers, usually 14-16 hour days. Children and women were employed as cheap labor, thus changing family roles. As industrialization took hold, it radically transformed existing social classes, empowering the land owners and the bourgeoisie. And this is crucial to our discussion. This new reconfiguration saw vast amounts of wealth remain at the top, while the lower classes were squeezed of it. Government gradually loosened their hold on industry, allowing it total freedom, in the laissez-faire spirit. Industrial tycoons controlled the market, exploited workers and grew rich. Poverty skyrocketed; the homeless lined the streets and the poor houses.

Yet, despite this, many saw this as a positive action because even though some suffered, society as a whole benefited. This ideal was encouraged by Jeremy Bentham and his philosophy of utilitarianism. Bentham postulated that humans were driven by desire and that humans should not suppress this desire, but embrace it, because it was natural. He believed that humans should pursue a path of "Social hedonism", because morality should result in pleasure. Basically, the "Principle of Utility" would have to be invoked, which asks the question: How many people will benefit? And it is this principle of utility that should drive all actions. Actions should only be measured by the way they promote the general good for the most amounts of people. Bentham's philosophy meshed perfectly with the new emergent bourgeois. Bentham's philosophy justified the social hedonism-and greed-of the new class. Industry also changed the landscape. Dingy, soot-encrusted towns sprang up, they almost resembled prisons. Life changed in a number of ways.

All in all, industrialization encouraged people to produce. What the common laborer produced was products for consumption. Laborers were no longer people, but simply units that could produce items, which could then be sold. And soon, this philosophy of production and consumption began to infiltrate society. I believe it produced a framework. The recently empowered bourgeoisie, the influential and powerful minority, manifested this philosophy of production and consumption. As Karl Marx lamented, the doctor, the teacher and shopkeeper all become wage earners, he insisted that the "chivalry" of these positions had been stripped from them; instead they had to be assimilated and subdued by industry.

Our modern day society still bears the imprint of industry. Since industrialization transformed America, especially during its infancy, industrialization left its mark. Later generations inherited the values and ideas from the effects of industrialization. It seems as if life has become a sort of assembly line where everything is produced to sell, whether it is religion or government. In many ways, Marx's prophecy still holds true, the "chivalry" of various occupations has been stripped and replaced with only the potential for production.

One area in which I can speak with certainty of is education. High School students today are not accessed on how they think, but simply what they know; they are treated as laborers, as ones who produce. Regurgitated facts on standardized tests are like shining trinkets on an assembly line. The students file into each classroom like an office. They are expected to produce As and Bs and 4.0s, as if true knowledge and creativity could every be measured by a number. While I understand the necessity of numerical grades, I do not understand the absolute reliance on them. Poetry and philosophy are hard to quantify and so students only take token courses. History is reduced to dates, math to numbers; both are stripped of their spirituality, abstractness and critical thinking. Like a foreman, the teacher drills his students to produce, not to think. These fact-filled but critically dead students are then sent out into the world, the world in all its grandeur, in all its sheer terrifying power bombard these helpless "adults." They turn to their industrial-education but are lost. Many are not equipped to face reality. Like broken old factory workers that are no longer strong enough to pull the rusty levels, they are discarded.

Industry has affected government and its policies in numerous ways. While many works have been written on this subject, I intend to focus on one: war. Today, I watch as the military advertises itself everywhere, it boasts a message for men (and now women) to do their patriotic duty. It calls on the some of the most profound issues dear to humans, pride, civic duty and honor. And for what? For exploitation! As my friend puts his life on the line in Iraq, as he does his "civic duty" for America, Halliburton is cashing in off his blood. As the rock Metallica band has so eloquently put it, these men are "disposable heroes." The military is a product, advertised and sold. And the benefits? The bourgeoisie upper class. This is by no means a new argument, indeed it was voiced since the revolutionary war. All wars have an economic component, but this is rarely if ever made known. I will discuss this shortly. Men and women serve to honor their country which is nothing but a factory decorated with an American flag.

Industry has affected other, more diverse and wide ranging parts of life as well. Youth sports no longer teach any morals, sportsmanship of love for the game. They only teach the youth how to quantity himself and make himself seem attractive to a buyer coach. Catholic churches pass around their baskets and ask for money to pay for nice new shiny stain glass windows while children in Africa are dying. Advertising has infected reality like a virus, everything is made to sell, everything is made to look shiny and new, but when you actually do buy, nothing works. Industry has quantified society, divided up into a lunch and cigarette break.

I wake up in the morning and the world comes to me like an assembly line. The billboards, the cheap gas, and the students I teach, everything. We are all made to buy, sell and trade. To quote the playwright Arthur Miller, "The only thing you got in this life is what you can sell." And if you can't sell anything, then you have to wait at the trof for something to eat. Society (and especially schools) preaches for citizens to be upstanding and virtuous. And yet, virtue doesn't sell. Virtue is an empty term that only the naïve adhere to while the rest makes money. Society tells its citizens to be virtuous, but it rewards the ones who sell, the ones who make themselves marketable from consumers[45]. Industry has been the most successful method of the last 500 years. It has unified society, induced all to labor, whether it is in an actual factory, or in their own life, it has told them (usually subconsciously) to labor for pieces of their own existence. When they finally save up enough, these pieces are hollow and they break.

---

[45] Some will undoubtedly point out an inherent contradiction in this statement. Virtue is not supposed to be "rewarded," that would defeat its purpose. But nonetheless, society uses virtue to manipulate ones who adhere to it, while rewarding ones who do not adhere to virtue. My point is that, while virtue should not expect a reward, unvirtuous actions should not be rewarded by a society that claims to be progressing toward order.

Industry is manifested by the majority because many benefit. Whether bourgeoisie, petty bourgeoisie or the principal of a "successful" school, someone stands to benefit. But the cost is too high. It erodes the foundation of society, it dulls our students, it weakens otherwise productive members to simply produce quantities. It sends young men and women off to die in a useless war. Rewards are given to the ones who can sell the most, not the ones who think. Industry has unified all, but in a detrimental, parasitical relationship. The ones at the top exploit the ones at the bottom while manipulating them into thinking they are doing their duty. Industry can be a means to better our society by providing goods that can help us realize order and progress. But it is abused as an end of wealth for some while exploiting the lower classes. And it is a hollow wealth that is not progressive. We have inherited a price, one in which may be too expensive to pay.

Reason, science, technology and industry have imprinted their face onto society. Yet they are all frauds, they are only means to an end. They are consummated in an unholy union; they dull our mind while pretending to prop us up. They are bulwarks against entropy. These defense mechanisms have evolved and have served their purpose-but like other mechanisms before them-only for some. Some can take temporary refuge behind their enticing goals. But industry produces empty things; the ones who profit receive only empty rewards. Perhaps I am too idealistic, but industry should produce goods to help order, not shackle the producers, and not turn everyone into a consumer.

In 2007 AD, another crucial issue confronts us, namely, the so-called "war on terror." While I do not mean to down play the very real threat of terrorism in our modern world, I do have a word of caution regarding it. I cannot fully ascribe to the "war on terror," simply because I think this notion can be abused and exploited. However, it is not the phrase so much, as it is people's fears. When people are afraid, they can be made to do anything. Our government has a waged a war on terror, I do not believe it is war that cannot end the threat.

Before I dispel the notion of the war on terror, I want to put forth a short history of Islamism and terror; however, it will not be the familiar candy-coated history that the government and the media have given us. And indeed, while I am offering only broad generalizations, I do believe they need to be seriously considered. Islam began in the 7th century with the teachings of the prophet Muhammad. His teachings united an otherwise disjointed area and formed an empire. By the 1300's this empire was at its height. The empire, as has been previously discussed with Christianity, became secular, almost by necessity. As it secularized, it met with a fierce resistance from fundamentalists, and gradually, many turned away from the secular. At the time of this inversion, the region to the Islamic lands west, Europe, was growing in strength. Eventually, by the 1600s, Europe was the most powerful region in world. By the late 1800s, European nations had carved up Africa and some Muslim lands for colonization. By the early 1900's, the last Muslim empire had fallen, leaving the Muslim world leaderless, confused and at the mercy of Europe. Unlike the west, many citizens in the Muslim world today do not enjoy a plethora material comforts or security; they do not have any effective bulwarks against the entropy. Entropy is a prevalent and intense as the desert heat. It is searing and painful, it creates a desperate longing. Entropy, with nil behind it, are the most powerful forces in the universe, indeed, they supersede the universe.

Terrorism is a threat to order. When the world trade center came down, it opened up a void or chaos. It was a desperate cry by ones who only know entropy. While there are various different reasons for one to be a terrorist, I do believe it is a capitulation. The lives of average citizens in Muslim countries are vastly different than average citizens in the west. Entropy is usually just a passing thought for many in the west and more well to do countries. But in a terrorist country, it is usually a way of life. Obviously, I do not condone terrorism. But I do not think bombing a country will make for lasting peace. I think aggressive military action will only prolong the conflict, and enabling the present system to endure. The education system in many Muslim countries is also vastly different than the American educations system. Many times, the humanities and the arts are not even taught, rather only the concrete, rigid math and sciences. While this is vast speculation, if one is not taught to critically think, how can they ever even begin to consider changing their world? Education and frustration, I believe, are the main contributory causes of terrorism in the modern world. Military action will only exacerbate the problem, contain it, but not remedy it. And I believe that in light of the contractual pledge that many make to live in a society and use its benefits, humanity has an obligation to aid and understand all its fellows, not bomb them (and I am referring to the terrorists and the war on terror alike). While this is a very lofty goal, I believe that it can be achieved through hard work, though the building of bulwarks against the entropy that encroaches on people less fortunate. Unfortunately is much easier to destroy something than to create and build it. Terrorism is a symptom of a much larger problem, namely that of displacement. I will discuss this problem in detail in a later section.

All in all, we live in a crucial time of absolute importance. The inheritance of the past four million years has been exposed. We have inherited contradicting and overlapping ideas. They tangle like barbwire over my skin. All is now exposed. All are means…but to what?

2007 AD and the Infinite

Society, at least in the west, is a volatile mixture of the last 3000 years. How are we, in 2007, supposed to progress with our overlapping inheritance? Before I answer this question, I would like to examine the state of the battle today. I look around, I see trees and people, and I think that all are fighting a battle, most of them unconsciously. I think a sort of stasis has emerged. I look at the faces of the people I pass on the street, I listen to my students, confused and garbled syntax proclaims nothing, just empty repeated slogans. A general apathy has spawned. No one knows what to think…how can they? I know my detractors will accuse me once again of being too esoteric, using too much poetic language etcetera. But it is the only way to describe what I see, what I feel. I speak to my peers, to humanity, and I honestly ask…what is it that you want? What do you live for? When you strip away all the facets of your life, what are you left with? After you vote, after you go to church, after you go to work, after you have sex, what is there but the void calling you, the inevitable decay of every cell and atom in your body? When you go to sleep, what do you dream about?

I fear that I have already become too esoteric in my assessment of modern America. Yet, a recent work, James Patterson's "Restless Giant" confirms some of my convictions. The book surveys America from Watergate in 1974 to 2001. In it, Patterson quotes Alexis de Tocqueville: "that strange melancholy which often times will haunt the inhabitants of democratic countries in the midst of their abundance, and from that disgust at life which sometimes seizes upon them in the midst of calm and easy circumstances." Patterson cites America as having this "strange melancholy" or restlessness even in the midst of its abundance and wealth. Is this not true? If one is to look, really examine an "average American," what would he find? I believe an astute observer may indeed glimpse this restlessness. While I am generalizing, I do believe my conjecture is valid in some respects. Whether a poverty stricken mother of four from a ghetto is surveyed, or a white collar CEO, I think in many, one would find this restlessness[46]. This restlessness would obviously manifest in different ways. Patterson cites the massive cocaine addiction during the 1980s, obsession over President Clinton's sexual liaisons, the LA race riots, and the rise in crime as forms of this restlessness in America. The question is, in a country as affluent as America, why do these problems exist? I am not naïve enough to think that everyone will be live together in a happy union, but nonetheless, if one surveys a newspaper, they will uncover forms of this restlessness, indeed, if one looks at their neighbors, they may find it. To echo Patterson, I think America in 2007 is a "Restless Giant." The source of this restlessness boils down to a need, an unquenchable need which government, religion, nationalism, industry, science, technology and materialism has not able to completely satisfy. Ultimately it is a need for order in the face of chaos. Yet it is not just a need of 2007, but rather a need that has been passed down and inherited since the creation of the universe. It is a desperate, blood-curdling scream which is drowned out by the revving of an engine, it is a scream that spans through space and time and all matter. A scream which many chose to ignore because it is too painful to answer it. But it must be faced, it must be asked.

I conjure up these points not to be fatalistic and prove that existence is meaningless, but rather quite the opposite. Our existence *now* is meaningless, because we all float in a lifeless stasis, pulled one way by some dying inheritance, then pushed in the opposite direction by another. We are told that order is everywhere, but really, it is far off and distant. We are being pulled apart one cell at a time. Someone once asked me how the world was going to end. I quoted T.S. Eliot and said: "The world will end not with a bang, but a whimper." And if we continue to exist in our present stasis, I believe it will happen. Various sections of the population will chase empty goals sold to them for the price of their existence, plasma televisions, and new cars, whatever. Millions will kill each other for Allah, oil or water. However, these things in themselves are not undesirable, and in fact, many are good. It is only our obsessions with them that hinder us, not the actual object. The world will gradually

---

[46] I do not believe all people would be "restless." Many would be more than content with their life.

work itself to death to produce a new car. If Jesus came back, there would be no one to save. We must recognize our present state. We must recognize that we can annihilate ourselves and any semblance of order. In the meantime we sell ourselves like prostitutes because that is what we have been taught to do by our boss-teachers. This is a warning. We are at an impasse, and we must fight to preserve ourselves. We must fight to realize the order that our ancestors craved. We have every tool at our disposal and we must use all of them, or chaos will rape us, use us to conceive of its victory, of the world without even the possibility of order.

In 2007 AD, we are a society of atomized individuals, linked together through bonds of commerce, lust and civil norms, but having no real connection. I am not original in this conjecture, I am echoing many 18th and 19th century German thinkers, but I have adapted it to modern society's situation. Yet, as an educator, I see this situation unfold in front of me. I see students with no real cohesion, just casual bonds. They produce only products for consumption. Material goods and pursuits have become blind alleyways. However, I do not propose a "Gnostic hatred of matter" or a disdain for the empirical. The empirical is the "stuff" of humanity. Indeed we are physical beings, but I do believe we have become distracted from our mental capabilities, indeed our responsibilities. What is needed, I believe, are "calls" to action. If man is left to his luxuries and vain pleasures (which again are not bad in themselves) then the modern stasis of society will perpetuate, and in words of Jean Paul Sartre, things will become "too weak to die."

America is a country built by force. Indeed it was forged in revolution, gained territories through wars and battles, and emerged as a world superpower after World War II, and then as the mega power after the Cold War. American dominance has been fostered through battle. This battle mentality has transpired across time and region. Businessmen, athletes, doctors, even teachers are competing for their own existence, because what else do they know? They have been taught to compete, taught to produce, that is what makes America great that is what makes it strong. Plato once remarked that an ideal civilization would be composed of warrior poets. Hardened by battle, but softened by poetry, they would be a perfect balance. America has only adapted half of this ideal. We are all warriors, fighting for some invisible goal. We compete and produce, this superpower has no time for poetry. But in the end, poetry may save it. Poetry is hard to quantify so it is forgotten. I do not contend that competition or production is bad in themselves, in fact they drive people to be better, and they give incentives. But when the goals are forgotten or manipulated or unattainable, the endless competition remains. People are driven like cattle, they are forced to wait in line for their own existence, and anything that cannot be counted is useless. America was built on force, built on labor, and so poetry is forgotten, useless because it cannot be sold or built into a foundation. But America cannot bury its poetry because that would strip it of its humanity[47].

---

[47] It is here I would to add a point. Some anti-Enlightenment thinkers stressed this point endlessly, Hamann and Jacobi in particular. They believed that if the Enlightenment ideals of universality, scientific reason and rationality were carried to an extreme, that they would imprison man in cold, mechanistic existence. These anti-enlightenment thinkers, while clinging to dogmatic faith, nonetheless raised crucial objections of science and reason. I believe some of these criticisms can be made of society today. Poetry and history have been regulated to the "dustbins," and only science and math are truly cultivated in students and the population general. The humanities must be cultivated and taught, man must learn he is not trapped in a cold mechanistic box, that he has

And yet, many of my polemics and accusations are not modern. They have been waged at various civilizations in various time periods before. However, they are still valid, still crucial. 2007 is a crucial juncture, but then again, every historical moment is a crucial juncture. Some junctures humanity has passed through for better, some for worse, and some by learning hard but valuable lessons. A great example of this the 1930s, 1940s, 1950s and 1960s. Specifically looking at America, society went thought massive upheavals in these decades, some of which contributed to the building of a progressive, open, enlightened society, yet others scarred America and left lasting prejudices and deep hatreds. During the 1930s, racism and discrimination racked America, not just toward African Americans, but toward all minorities and even women. Most historians would cite that the United States failed to live up to its ideals in the 1930s and 1940s, that all men were not created equal. And I wholeheartedly agree with that assertion. I believe America failed, but, in its failure it forged the way for future generations. Many failures of the 1930s and 1940s were challenged in the 1960s. Not all were rectified, even today we still live with the failures of our grandparents, but overall, I think almost all would agree that positive results did occur and change society for the better. I bring up this point because I think it relevant. I think it illustrates the capabilities of social change, even in the face of failures.

2007 AD is a crucial historical moment, fraught with the burden of past failures and contradictions. Society must either rise to the challenge of advancing the moment, more specifically; society must realize if it wants order it will not come without effort. Either society must work for this, or remain in a terminal and fatal stasis, declining, awaiting the end of order, awaiting the end of the world[48]. But not some cataclysmic, apocalyptic end, but rather a piecemeal end, an anticlimactic end, limping on in fragments. The world will not with a bang, but a whimper.

Arthur Schlesinger, in 1949, wrote a piece called "" which signaled the advent of the "new left" in America. While I do not agree with many of his points, I do agree with his conception of the "mass man." Schlesinger contended that as a result of historical processes, since the end of the middle ages, man had been stripped of his familial associations, such as church, family, village etcetera. As a result, modern man is an empty, hollow entity with no real allegiances, which he calls "mass man," And it is this mass man that is easily swayed by the likes of totalitarianism, communism, Hitler and Stalin. In 1960s, Schlesinger and the new left equated mass man with materialism, and an apathetic society more bent on personal gains that political activism and change. I would like to extend Schlesinger's idea. I believe that "mass man" is the result of the historical process outlined above. But mass man is not an inevitable result, but rather the result of the failures of historical processes, and the subsequent contradictions inherited by future generations. I believe "mass man," in modern society, is a bloody stump, a legless, armless, eyeless stump of a man, ground down and hacked to pieces by failures and contradictions. Yet, mass man is injected with "historical and cultural" morphine, such as dead ideological slogans and selfish, self-preserving methods of materialism. All the while "man" is alive, but a

---

the power to transcend this artificial imprisonment with his own imagination, an imagination that has been stifled, deadened and sold for profit.

[48] This is a condition that some desire. Indeed, order is not a universal desire.

better term would be "surviving." Yet, these stumps[49] still have brains, brains blunted and stifled, but able to be rehabilitated.

So the question that remains is how to rise up to the challenges of 2007? How do we rise above the contradictions of our distorted inheritance? How do we answer the call, indeed, how do make the call? The answer, I believe, will shock many, and turn any reluctant supporters of mine completely away with disgust. My proposition will undoubtedly be viewed by many as too idealistic, poetic, or esoteric for practical use in a society like ours. However, I will try to argue for a practical usage and attainment of this lofty goal and application to society.

I believe the human intellect already is in possession of the means to its survival in the battle. It is something that the intellect has possessed in some form or another for centuries. It is the conception of *the infinite*. However, must simply view this conception as transcendent and unattainable. Most people move through life like a lump of dung, clinging to whatever it happens to fall on. They are the bloody, legless, armless stumps. But it is the infinite that can rehabilitate them. The conception of the infinite is usually wrapped up in a vision of God and we have been told for centuries that our finite bodies are not equipped or not holy enough to handle this vision. However, it is our finite minds that *produced* this conception of infinity. I believe infinity can be "the call" for man to heed his mental capabilities, and it is through upbringing and education that this call can be administered, which I will discuss shortly.

It will be necessary to attempt to define the "infinite." Yet, it is an extremely elusive term that can accommodate or defy any definition given to it, simply because there is no one true definition. I do not propose the infinite is not some linear goal of humanity. Rather there are infinite infinites that can all exist simultaneously. Different portions of humanity may reach different infinites at different times. I propose just a handful, but these are by no means exclusive of any other means of infinity.

A crucial infinite is infinite reality. What I mean by this is one when one looks at reality, at the trees and roads they have seen everyday, they must re-look at them, un-focus their vision. One must see the outlines of the road detached from the road; one must recognize that those lines then extend infinitely. All around one are lines embedded into objects, even embedded in themselves. Objects end in space, they are finite, but the ideas that are spawned from them (i.e. lines and numbers) can become infinite if looked at correctly. One must realize that bodies and objects can be defined as ratios and numbers[50], parts blood, steel, glass and these numbers extend infinitely. I can be represented as two arms, two eyes, or a billion cells. And all of these numerical representations are subsumed into the infinite because numbers are simply human markers, human delineations that extend infinitely. These pockets of infinity abound throughout reality, and they can be the first contact one has with the infinite. The human contributions that a mind makes to reality already contain the infinite. Time and Space may not be inherent in nature, just in our own minds. Instead of rigidly subduing reality and forcing it to fit into time and space, one can contribute the infinite to reality.

---

[49] I do not think all are like this, in fact there are gradations. Some people are completely blunted, unable to think for themselves, apathetic to the world. Others have limited cares and abilities to produce change.

[50] Aristotle, Metaphysics

On a more practical level, I believe another crucial infinite is that of infinite religion. As I have argued in my historical analysis, religions have failed to bring man to the infinite; however, their failure was not in vain by any means. Religion brought man in contact with the infinite. Yet, with their failures, I believe it is time to subsume all religions together. I do not call for one monolithic religion; in fact I call for the opposite. People should worship their own gods, but at the same time, realize the legitimacy of other religions. Instead of competing, compromise. This will anger many die hard adherents, but really, has any religion ever really liberated their people? Almost every religion has betrayed its own values. I think Catholics should realize the worth of their own religion as well as realizing the worth of Buddhism, Judaism and Islam. All religious adherents must realize that other systems of thought have meaningful contributions to make and not one system should be held against an other. All have positive ideas to contribute and all have failed in various ways. Yet, many refuse to admit this and instead of realizing harmony, many adherents remained locked in a prison of their own history and dogma, believing that their religion, their method of apprehending the infinite is better, while all others are sinners.

The power of poetry is something I have mentioned throughout this work. Yet, include it in this section because I believe poetry is a lost and forgotten art. Yet, it is more than an aesthetic art; poetry is a method to apprehending the infinite. Indeed, poetry is the language of the infinite. Logic, and reason are only parts of the whole, and I believe poetry is their complement. Children must learn to communicate in poetry as well as math and grammar. In the book by Sid Barrett "The Irrational man," Barrett claims that poetry is the channel to the irrational; the poetry is a link with this irrationality of our own lives, with the void behind all. Poetry cannot be quantified by reason, and thus it has the power to transcend the mind, transcend the ordered world.

Infinity can be whatever anyone wants it to be. Infinity is the antithesis to Shelsginers "mass man." Mass man is a man stripped of all allegiances and motivations, and this nihilistic conception is partially justified because man has seen everything he and his ancestors had believed fail. But from these failures, with the infinite, man can create a new existentialist path-if he so chooses. He can buttress the inevitable decay of all existence-if he so chooses. The choice is for man and man alone. Infinity is a terrifying pleasure where opposites mix and new things can be unborn. Infinity is the key for our release from our contradictory, nihilistic existence. It is here where I will reintroduce Kant. The human mind, once in possession and control of infinity, must then *impregnate* reality with it, must infuse reality with the infinite, and must transplant the infinite blood into the stale veins of its life. Hegel's goal was to for the mind to bridge the gap between what the mind sees, and what is independent of the mind. While this may be an esoteric or even spiritual endeavor, I think it has some merit, along with Kant's assertion. The human mind, once educated of its possibilities, namely the infinite, must then structure reality accordingly. Instead, it cowers in the face of the infinite and builds reality with what it knows. Thus, the mind remains disjointed and alone, and cannot build a reality in which it is truly free.

I believe the ultimate infinite is infinite awareness. Humanity must be made aware of the battle that rages within their every cell, in every atom of their every possession. Humanity should not live in metaphorical prison cells. The wardens are invisible, whether they are overbearing advertisements or inherited contradictions.

Humanity must be made aware of the situation. When I turn on the television, I should not watch dozens of news stories about hotel heiresses. In what way does that contribute to humanity? I am saying that pleasure and relaxing activities should be disbanded; on the contrary, I think these notions are extremely apt at giving humanity momentary reprieves from the stresses of daily life. Yet, humanity can not live in a state of perpetual reprieve; humanity must pull up their boot straps, sift through the contradictions, and pursue the higher mode of life which is in their capability to achieve.

With infinity, humanity can bear the fruit of millions of years of defense against entropy. With infinity, humanity (even in itself just a defense mechanism against entropy) can envision a place without the possibility of pain and suffering, of eternal and divine achievement.[51] Within infinity, the things that regulate our existence like prison wardens would vanish, or could be bended to suit our imagination. The space between wife and husband would become an entity in itself, an animal divided by the contours of the world, time would move to wherever we wanted it to move. Infinity is as of yet untamed, uncontrollable, but nonetheless, it exists, and we have created it. I am allied closely with certain streams of existentialist philosophy when dealing with the infinite. Existentialist philosophy has at its heart the concept of raw existence. It places existence before essence; we are born with existence and create our own essences, rather than being predetermined to an essence before birth. Man creates his own essence, his own definitions to live by. Indeed the only thing in which man is born with is the ability of choice. He can choose his own destiny. There is no church guiding him, no universal standards of law or reason, only choice. There are no eternal forms to aspire to; there is whatever the mind wants to create.

Thus existentialism expounds the creative principle in man, it places the highest emphasis on mans ability to create his own existence. I want to extend this existential conception of freedom to my conception of the infinite. Man can create systems of order to help him survive and progress, but these systems are artificial, or at least incomplete. Man has the ability to endlessly create whatever he wants, as evidenced by the conception of infinity.

It has been brought to my attention that certain remnants of my philosophical thought can be construed as a fascist sense. Such as my privileging certain members of society, namely productive ones, and ones who can apprehend the infinite. In a sense, this assertion has some validity. I do promote a type of elitist vision of society. But my elitist vision is not based on some arbitrary standard such as color or race, instead, my "fascism" is based on talent, skill, imagination and above all, hard work. When a society refuses rights to a certain portion of its population based on irrelevant qualities, it stifles the individual, but more importantly it hurts itself by denying itself a potential contributing member. Citizens of a society cannot make accurate judgments when portions of their population are ignored. And so, I espouse a "meritocracy," in which even the smallest contributions to society would be greatly valued. Only the

---

[51] Note the usage of the word achievement. Our success will require effort, nothing will be given, and unfortunately, many will not put forth the effort.

loafers, shirkers, imposters and charlatans would be discriminated against because they refuse to contribute, or refuse to contribute to the common good[52].

Once again, I must stress the crucial importance of the present juncture. Society is truly at a reckoning point. With the technological advances that have emerged, namely, the ability to annihilate ourselves, coupled with all the contradictions of reason, religion, politics and industry, society must not succumb to chaos. Society must resign itself to a broken existence or overcome it. However, this truly is a battle; there is no Hegelian dialect that presumes order will triumph. I think if the current situation does not remedy itself, society will overlap and the competing facets will destroy each other, they will cancel each other out in a fraction. Factories will produce billions products for no one, but the factories won't stop; the assembly lines will run until there is no more room in the world. We must tap infinity, bend it like a vein and feed our hearts, and we must be taught how to do this from our birth.

Infinity is a bridge of sorts. Man has created something that will outlive him, which will outlive everything. Truly then, what is infinity? We can glimpse it by trying to conceive of the last number, or we can imagine a parallel line extending infinitely throughout the universe. But these are scientific-mathematical conceptions. What I desire is a more tangible entity. Infinity is the ultimate source of creation. Within the infinite, to quote Edmund Burke, all opposites mix. However, I believe that the human mind can take infinity a step further. Within the infinite, I think the human mind can create new boundaries, infinite boundaries, which would then cease to be boundaries.

Man has had a conception of the infinite for ages. It was vested in many of the classical gods, and in other parts of the universe and the heavens. However, with this came a false sense of security. Man thought he knew infinity, man made sacrifices to please it, but man never had control. Infinity lurked behind everything. Infinity has always been at the heart of the battle. Order needs infinity to conceive of a world with no possibility of nil, and chaos needs infinity to eliminate the possibility of order. Infinity is pure imagination, imagination unbounded. If man could harness this awesome power, he would "evolve" from *homo stasis*, to *homo infinitus*. And this would be an evolution at man's disposal and control; it would be evolution of man, but directed by man in any direction he chose. Truly, with man's technological advances, he could "evolve" anyway that he wanted to.

What is man? Man in himself is a self defense against entropy. "Man" is an evolved animal. Previously, I had stated that man's intellect evolved as a weapon against chaos. When "man" first used a rock as a tool, or first cooked his food, the battle changed. His intellect became the strongest bulwark against entropy. But the intellect evolved as well. It initially used reason and logically thought out problems of daily life such as food and shelter, but then it turned its reason to questions of the universe. Perhaps infinity is the evolution of man's intellect into a higher state. But, once again, it is not a Hegelian evolution or teleos. Infinity will not bow down to order, it must be taken. But how? How does man straighten out the pieces of his broken, contradictory existence and aim for the infinite?

I believe one way in which society can remedy this is through education. As I had previously mentioned, the youngest members of our society are routinely

---

[52] For example, handicapped people and mentally ill people would be aided by other, more self-sufficient members of the population. A cultivation of contribution would occur.

regimented and browbeaten with an industrial type education. Children are forced onto a metaphorical assembly line, in Marx's words; they truly become appendages of the machine. From kindergarten onward, children are not taught to think, only to produce. But how can children even begin to apprehend the infinite if they are only taught to produce test scores? Test scores fall miserably short in the face of infinity. Teachers must teach their children that there is a greater reality than a test score. Education must begin and not end with a final exam. History is not dates, it is blood and ideas, math is infinity and negative numbers, and teachers must expose the infinite within reality. The hiring of teachers should be of the utmost importance in any society. How can any society improve itself without educating its young?

Too many times, teachers are ridiculed. It is an un-respected profession, and unfortunately while some people do become teacher to just "get the summers off," many teachers do genuinely care for their students. However, the teachers have been trained in an industry-type of fashion as well, so they are helpless in educating their students. They do not truly educate their students; they merely resuscitate them everyday to a dying reality and lifeless version of their existence, a version which can only succumb to chaos because it refuses to pursue order. I do not mean to say that all students would magically be enamored with a great teacher. Unfortunately many would still succumb to chaos even if the opportunity to pursue order were laid at their feet.

I believe that students, who do not contribute, should be removed from the classroom. As an alternative, these children should be made to contribute. The best option for these children is probably some type of vocational training. Students-who then grow into adults-who do not contribute to society, pose the biggest threat to society's advancement. For one thing, the individuals' do not contribute, and usually other productive individuals are then forced to take care of the unproductive individual, thus severely compromising their abilities. I believe these individuals must be curbed, their idleness is a cancer, and it threatens to destroy the order which has been fought for. And this idleness is tolerated in the present education system. These children are play-cated, their idleness is attributed to some crutch such as Attention Deficit Disorder.[53] Children are given endless chances, and still, they produce nothing, except problems for everyone else. These children then enter society and produce nothing but expect society to give back to them. All members need to be productive. This is the only inconsistency with industrial education. In a factory setting, unproductive workers would not be tolerated, yet in the classroom, they are coddled. Ironically, I think this is the one aspect of the industrial inheritance that should be incorporated.

Education is not strictly limited to the classroom. Classroom education is formal education. In a modern society, almost all are bombarded with informal education. Whether is advertising or unwritten social norms, a collective consciousness is manifested (however, this consciousness is derived from the contradicted and overlapping inheritance). We are bombarded with images of what is supposed to represent order, but when logically examined, represent anything but. One commercial is for diet pills, the next two are for Papa Johns Pizza and Kentucky Fried Chicken. How can society realize progress when the messages that are sent to it contradict

---

[53] I do believe this condition is real and serious, but not to the extent that it is claimed in schools today.

each other? Children are told to work hard and do well in school, to play fair and always do the right thing, yet they see illiterate, immoral athletes paraded around in front of them. What people are told, and what people see contradict each other. Society cannot be built on a contradiction, let along progress from one. Teachers and philosophers that earnestly seek the truth, that ardently desire a better, or at least a truer picture of life for humanity, are discarded as mystics and idealists. Instead pop stars who shave their head make news, like her contributions to society have affected any, save for presenting a false image of beauty for young girls to adhere to through bulimia. How can humanity be expected to achieve anything beyond their own stomachs or loins when these manifestations are imprinted on us? How can I be expected to hold anything sacred when all the things I am supposed to hold as sacred are prostituted for a new stain glass window, or to make Halliburton's stock raise a point? Criminal-athletes have the nerve to appear on national television and demand more money while homeless bums starve in the gutters. I do not propose a communist society where money would be divided up equally, and personality is discouraged, but, is it necessary for this man to publicly state that 12 million dollars is not enough for catching a football, while it would take me roughly two decades to acquire that much money?

If humanity is to recognize the infinite capabilities which it has created, and has possessed for centuries, certain institutions must be radically transformed, reformed or abolished. Education-both formal and informal must be radically reformed. Students must become students of reality, of infinity. They must learn that they posses the font of all imagination, that if they truly wanted to, with an inordinate amount of hard work, they could change the society they live in. They have the power to imagine anything, even things that they do not know yet. That is the beauty of infinity. It is mystical, but logical at the same time. It exists everywhere. I propose that society must replace industrial education with infinite education. Students (and teachers) must not be taught to produce for a shop window, they must be told not to sell their blood to the highest bidder, but rather to realize its infinite worth. Order has always craved a vision of itself with no possibility of nil. For millions of years order has produced entities to combat chaos, all of which have failed to be an end, but which can now serve as a means to a greater end. Government gave us order and structure, religion gave us ethics and the infinite, reason gave us a rational judge able to weigh consequences and make better decisions, science improved our existence, nationalism gave people a rallying point and industry made more items accessible to more people. The infinite is the bridge between all these things-if used right. If used wrong, or if left idle, infinity will usurp all of orders' achievements, and turn them against each other in a sort of "civil war." If order is left in its present state, it will devour itself like cancer until nothing remains, not even the possibility of order.

However, education has to start in the home; it has to start at birth. Children must be taught by their parents that they possess the infinite inside them; it is a parent's job to actualize the potential in their children. But the parent cannot simply do this for child, the parent must show the child how to realize the infinite, not reach it for him. Parents must read intelligent books to their children; they must limit video games and television. Parents must immerse their children in activities of higher learning. A child is a finite vessel that contains infinity. Unfortunately, most parents raise their children only to be consumers and not creators. They are taught to buy their existence piecemeal, they are not taught how to put forth effort, they are not taught how to achieve anything, simply buy it, they are not taught how to create, only to consume. A parent must teach the child that they are harbingers of creation, that they all have capabilities to aid society. They must teach them that a battle rages between order and chaos and they are at the center of it, that they are the greatest weapon. They must teach their children that order is not given, and that they all posses an ability to contribute to its preservation. And then, when the child is old enough, he then will go to school, and his infinite education will continue.

I understand this to be an idealistic vision. Socio-economic or behavioral factors many times limit a parents or teachers ability to discipline or provide for their children. Personal differences may compromise a student and teachers ability to communicate. Also, another major obstacle to infinite education is personal ability or talent. Not all members of a society can realistically be expected to contribute the same. However, what I endeavor is that all members of society contribute to their abilities and circumstances. A person of mild intelligence can still contribute. He should not be forced to fit into the mold of another child, a child who is exalted because he has achieved a 4.0. Some children will never receive 4.0's; some children will never be able to play football. Society subconsciously sends a message that these children have nothing to sell. While these children are told as youngsters that they can achieve anything and they are infinity special, this is a hollow message. If they cannot produce anything to sell, anything to display such as touchdowns or grades, they are silently ostracized. They become second class citizens whose only hope of any success is to fight for what the first class citizens throw away. As these children grow up, it is their parents and then teacher's job to show these children posses the infinite, that their creative, critical abilities are not fantasy. Just because one cannot quantify imagination and sell it, doesn't mean it's worthless. Poetry is logic, and can be used to understand the universe. Poetry is the logic of existence. And yet, since it cannot be sold or put into a shop window, it is not valued anymore. And I believe it is the role of the parent and then the teacher, to make the child see that the 4.0, while important, is not an end. The 4.0 is a means to an end. The 4.0 is the beginning not the end of education. The proscribed end however, does not have to be an end, it is infinite, and can be what ever the creator (not the producer) desires it to be. Children must be made to realize that after 13 years of elementary and secondary school, after 4 years of college, and then masters and doctoral degrees, all that knowledge is only a beginning, a gateway, a spring board into whatever they can create.

All must contribute something, either intellectually or physically. All must be taught of their capability. If their capabilities do not conform to the majority's manifestation of sanity or normalcy, many of these children (who grow to adulthood) will be unable to contribute to society positively, and at this point, the infinite which they possess will become dead to them, and another weapon will be lost[54]. Once

---

[54] I would like to venture a point made by Hegel to bolster my own claims. Hegel contended that if a society failed to move past its social stage into its political stage, if members of a society failed to harness the political in them, the infinite or namely the ability to move past their own materialism, Hegel believed they would degenerate into "bad infinity," or a state in which people simply craved insatiable material demands (akin to the Buddha's conception of cravings).

again, people will point to the fact that some members of society may just be incorrigible and unable to integrate and participate. Or some members may simply just not be able to comprehend the infinite. This unfortunately is absolutely true. Not all members will pursue order. It will take the effort of other productive members of society to correct and deal with the non-contributing members (i.e. police officers, FBI). What I propose is simply that members of society, despite their differing levels of ability, must realize some cohesion, they must look and think beyond their simple world and realize the order they can work toward.

While many would remain incorrigible, I do believe that many potential non-contributing members of society may be significantly helped through education. But education must be formal and informal, society must be saturated. I believe at this crucial juncture in time, where humanity sits a top a throne made of broken and contradictory inheritances, the only thing which can negate and verify all the contradictions is the infinite. The infinite can bear fruit to all of the contradictions and tangled barbwire that we have inherited, the infinite, humanity's own creation, can heal every wound. Within the infinite, man can keep forging more elaborate patterns of order to fend off the chaos which looms everywhere.

At this juncture I would like to make a distinction between children and adults. The older a person gets, the harder it is for that person to adapt new ways or break old habits. This is extremely true of education. It would be hard for many adults to learn the infinite. However, I do not feel that adults are totally lost and should be ignored. Adults are the key. They might not be whole-heartedly able to adopt the infinite, but they are the ones who have to pave the way for the children. They have to believe that the children can adopt the infinite. Because if no one teaches the children, how will they learn? How can anything ever change? I am not venturing that teachers should change all their lesson plans, all I venture is that teachers should not simply "teach to the test," or just teach cold facts, instead, all I venture is that teachers should challenge their students, that along with teaching required facts (which are important), teachers also need to teach their kids how to think, how to use those facts, teachers need to teach their kids how to create.

Not everyone will be able to grasp the infinite, either adult or children. And some will be more receptive to the infinite than others, some will only grasp it superficially, some will grasp it wholly. It is not a black and white occurrence, but rather one that will take place in gradations. I do not believe that there will be some mass conversion where everyone will accept the idea of the infinite. Rather, if it does happen, it will be a slow and laborious process, affecting different people differently. Yet, this is how change occurs, slow and laboriously. If the infinite is grasped, then little by little, humanity will change; humanity will slowly realize its own potential. If enough people embody the infinite within, the infinite created by their ancestors, if people realize they have the ability to create anything they want, humanity will be victorious in its battle against chaos and ultimately nothingness.

In summation, the infinite is a concept possessed by humanity, or more accurately, a conception created by humanity. Man can reach infinity through poetry, through philosophy, through music and through education. Within infinity, mathematics is more than numbers, history more than dates, English more than grammar, education more than grades. All becomes all, all becomes none, all becomes true, and all becomes false. Nothing is for sale, just for understanding.

Examination of American Society in 2007 AD

While I do not want to portray America as the only important country in the world, it is by far the most influential. And so, I feel that I must dissect American society and examine it to better understand the world. In the beginning of this work I stated that I did not want to solely concentrate on American history of the west. Yet I truly believe America has the potential to become an ultimate bulwark, yet it is floundering. Indeed, I am not original in this conjecture. Many thinkers have hailed America as a sort of Promised Land, or "city on the hill." Some might even label me a propagandist. However, I think America, at least theoretically, offers its citizens an amazing amount of freedom. While many would undoubtedly argue with me on this point, (even my own words contradict this at times), what I aim at is to uncover the principles that America was founded on. I believe *potential* America, not *actual* America offers this opportunity. I believe that it is those principles which, if adhered to, could help lead society to the infinite. So, I believe a discussion of American in its present state must be examined. Much like an alcoholic, America's faults and hypocrisies must be exposed, America must admit to them and then move on to realize the infinite.

I believe that American society is divided into three large classes. Each of these classes are then divided and subdivided.

> Neo-mercantilist class
>> Oligarchy-elect
>>> US Government
>> Oligarchy
>>> Large corporations, lobbyists

> Citizen Class
>> Middle Class
>>> Mainly but not limited to affluent whites
>> Lower Class
>>> Mainly but not limited to minorities

> Fringe Class
>> Homeless, insane, convicts, elderly, illegal aliens, abjectly poor

American society is not a true democracy. Rather, it is an oligarchy. Only 548 people have the authority to propose, pass, veto and amend laws: the president, the

Supreme Court, the senate and the House of Representatives. However, it is quite possibly the best oligarchy in the world. It assures its citizens a relatively high level of national and domestic security, numerous civil liberties, as well as more than adequate heath care (for most) and technology[55].

A symbiotic relationship must form between the levels in order to allow the system to operate. And it is this symbiotic relationship, one that benefits both parties (but not equally), that is the crux of our government. The neo-mercantilist class has assumed the political as well as economic power over its constituents, which are intertwined (hence the name mercantile). The oligarchy-elect subclass creates the laws and policies. However, it is the oligarchy subclass that draws its strength from the government, and sometimes guides its policy. A sizeable minority of business elites and corporate leaders, lobbyists and billionaires have a heavy hand in directing the domestic and foreign policy of the United States.

To illustrate this symbiotic relationship, the best example is to study the classes during times of war. Wars are usually rooted in ideology, where all classes join together to achieve a common goal. Beyond the rhetoric of war time patriotism, there are material things to be gained, and almost always by the neo-mercantilist class. And I believe this is the basis of war policy. Despite ideology, no powerful country in a position of leadership, will fight a war unless there is something to be gained. What would be the purpose of mobilizing millions of men and women, of spending time and money coordinating the war effort, if there was not material possession to be gained-at least for some? The neo-mercantilist class mobilizes the citizen class to fight its battles, and promises it various things in order to achieve its own material goals. Thus, the symbiotic relationship is born.

The Revolutionary War, which is touted as a war of freedom, if examined closely, is a much more complex situation. While it undoubtedly was a war to throw off the yoke of British oppression, it was not universally accepted by the colonists. In fact, only 2 out of 5 colonists were true "patriots," 2 out of 5 were "loyalists," and 1 out 5 were undecided. Hardly the romantic episode it has been painted to be, the revolutionary war was mainly led by wealthy and influential colonists who had sensed that Britain's increased interest in the colonies after the seven years war could hurt their business[56]. The neo-mercantilist class mobilized a vast majority of the citizen class, and even slaves in the fringe class to fight the economic oppression of the British. I am not contending that a British threat didn't exist. All I contend is that the revolutionary war was not a romantic struggle led by an oppressed people, rather it was a struggle led by a minority who sensed they could be hurt economically, and this minority then extended this economic domination to all facets of society and mobilized lower rungs of society to fight. As T.H. Breen argues in his book "The Marketplace of Revolution," the common glue that held the colonists together, from New Hampshire to Georgia, was a bond of materialism. As a plethora of British imports flooded into the colonies, more and more people from all classes were included in society, because as

---

[55] I know many do not have these things, but the vast majority of Americans do.
[56] The class system I have outlined above was not the same, but roughly the same divisions existed. The neo-mercantilist class was composed of the oligarchy who decided the laws, and the oligarchy-elect composed mainly of businesses men who helped to guide policy. The middle class was made up of the farmers, while the fringe class was made up of the same, except with the addition of black slaves.

the goods became cheaper, more had access to them. After the Seven Years War, the British tried to increase their hold on the colonies. As stated before, many opposed this new British intervention. Many began to feel that they were inferior to the British, when all along they had assumed themselves equal subjects in the empire. While the revolution was a complex event with a myriad of factors, political, social and economic, in the end, it was economics that fueled the colonists discontent. As Gary Nash argues in his work "Urban Seaports: Crucible of Revolution," The lower classes in the colonies had been downtrodden and economically squeezed for almost a century. They resented the upper classes as they drove by in their luxurious horse drawn carts while they, the owe classes, couldn't afford to feed their families. Nash argues that by the late 18th century, this consternation had been re-channeled against the British, who were now billed as the oppressors. I am not implying that the lower classes were simply had or duped into fighting, but nonetheless, the lower classes did turn to many of the upper and middle classes for leadership and organization, and many in the upper and middle classes may have not have always acted according to the professed patriotic rhetoric, but to a more hidden agenda of self-interest.

John Hancock, signer of the Declaration of Independence, a sterling patriot in the eyes of all Americans, was in actuality an illegal rum-runner who tired of the British harassing his smuggling business-which they had every right to do. When the war ended, many ordinary men, patriots who fought the evil oppressive British, came home only to find their lands confiscated for non-payment of debts. Patriots who fought were kicked off their land. Daniel Shay organized a rag-tag rebellion of patriots to fight the federal government that had evicted them. Shay's rebellion was crushed, but the significance is great. This could not have been the America they fought for, but alas, once the citizen class was no longer needed, they were exploited, taxed and evicted. For many of the lower classes, the revolution had not changed much, had not alleviated their suffering. The revolution did not alleviate the suffering of Daniel Shay. The agitation caused by Shay's rebellion was one factor in the reevaluation of the Articles of Confederation. These examples illustrate some of the inconsistencies of the romantic vision of the American Revolution.

The War of 1812, a largely forgotten war, is sometimes referred to as the second war of independence. The main cause of the war was the British impressments of American sailors, as well as the continued harassment of American ships by the British military, and the presence of British forts in Canada. War Hawks in congress declared war on Britain (who were tied down with Napoleon at the time) to stop the impressments of sailors. This seemed like a valiant cause and many supported it. Yet, a group of Northern shipping merchants almost broke away from the union in the Hartford convention. They did not want to go to war-even though the war was supposedly to help the beleaguered American shipping industry. The claimed that a war would destroy their businesses, much like the embargo acts of Jefferson in 1807 and 1809. Instead, they accused the "war hawks" in congress of declaring war not to aid the shipping industry, but rather as a pretext to seize Canadian land with its lucrative fur trading business. After declaring war, the United States, largely unprepared, invaded Canada was duly defeated by British forces. The rest of the war was completely defensive, with the United States stalemating the British. The actions of the neo-mercantilist class were barely disguisable. The faction most affected by

British impressments of sailors threatened to leave the union if the war was fought. But the neo-mercantilist mobilized the citizen class and fought the war anyway on that very pretext.

The American Civil War has been debated and hashed for decades. At the outbreak of the war, the United States was a young nation, not even one hundred years old. The Neo-mercantilist class realized the detrimental effects southern succession. It would weaken the fragile United States. And so, the Federal government, on the pretext of slavery, fought a bloody war to preserve the union. I am not contending that slavery had nothing to with the war; I am not even saying that slavery was not a major cause. However, there were other, less visible, but nonetheless crucial factors. The Neo-mercantilist class realized that if the south succeeded, the north-the federal government-would lose the trade of the Mississippi river, which at the time was the main avenue of trade, and northern businessmen would lose the lucrative profits from the southern cotton industry tariffs. These economic causes undoubtedly contributed to the declaration of war, the neo-mercantilist class saw its economic interests threatened and mobilized the citizen class to fight.

The Spanish American war was supposedly fought for the battleship Maine, after Spanish forces sank it in the harbor in Cuba. After this, the United States declared war on Spain and "liberated" Cuba. But the explosion was never proved to be committed by the Spanish. In fact, many believe it to be a faulty engine. Nonetheless, the United States defeated the Spanish and took control of Cuba, as well as the Philippines, which were both coveted strategic areas. The United States even inserted a clause in the newly independent Cuban constitution, called the Platt amendment, which forbid Cuba, a free country, to make treaties with out American approval. The neo-mercantilist saw a Cuba as a crucial gain and mobilized the citizen class to fight for it. Even to this day, Guatenamo bay is a reminder of Spanish American War.

World War I was a very different war than any other up until that time. The United States fought a European enemy on European soil. World War II saw the United States do the same thing. The reasons for America's involvement in both wars were to defend the world against tyranny and oppression, especially World War II. Propaganda posters and leaflets instructed the public of their duty as Americans to fight. However, many prominent businessmen had vast amounts of money invested in the Allied powers, in both wars. If the Germans had won, millions would have been lost. America was also in the throes of the Great Depression, and the war was an easy way to get out of it.

I do not mean to degrade America, or its values. I do not mean to degrade the reasons for entering these various wars. All I ask is for a deeper examination of the causes of these wars. There was more than ideology and rhetoric at stake. In every war there were real, tangible goals to be won, but not by the citizen class, they were only privy to ideology. It was the neo-mercantilist class that would receive the benefits by mobilizing the citizen class. Almost every single conflict has been a "rich man's war and a poor man's fight." The citizen class is mobilized with ideology to fight for the neo-mercantilist economic goals.

Today, as the war in Iraq rages, patriotic commercials abound on television, and as American soldiers are dying, rich executives from Halliburton line their pockets.

The citizen class had been told that Iraq and Afghanistan were allied with each other, they posed a joint threat, Sadaam and Bin Laden wanted to destroy the United States, which is true. However, they also wanted to destroy each other. A major motivation for Bin Laden's attack on the US was his resentment over his native Saudi Arabia requesting assistance against Sadamm Hussein in 1991, during the first Gulf War. Yet this fact is never mentioned, neither is the fact the United States supplied Iraq and Afghanistan with weapons to fight to the Soviets and the Iranians during the cold war, weapons which later killed Americans. After the first Gulf War, conditions in Iraq were terrible; however, this was mainly due to the sanctions that the United States had placed on them, mainly prohibiting them from selling oil on the world market[57]. Sadamm remained rich, while the country starved. Sadamm had no part in the World Trade Center, he also had no weapons of mass destruction. The United States fought Iraq on these pretexts, when they turned out to be false, the United States than changed its policy, and instead fought to "save the Iraqi people from a dictator, while in many other areas of the world, such as North Korea and China, authoritarian governments oppress their people and commit flagrant civil rights abuses, yet the United States does nothing[58]. The Neo-mercantilist class has mobilized the citizen class to fight its war and make it rich, and it has brainwashed them with ideology.

The most crucial class is by far the citizen class. This class has developed a symbiotic relationship with the neo-mercantilist class. As long as the neo-mercantilist provides the citizen class with domestic and foreign security, comfort, materials, technology, the citizen class will remain docile and not act against the neo-mercantilist class. Coupled with this symbiotic relationship, is an almost insatiable need of people to emulate and adore others, specifically those in a position of power. This "personalist paradigm" has been evident throughout history. It is almost as is people in the lower classes, not even poor, just of lower classes, need a type of "role model," whether it be a government official, a news reporter, or a pop culture icon. As if it is easier to emulate an already established figure and regurgitate their views, regardless of their talent or stance, than to realize the power and potential in one's self. Pop culture icons, celebrities, artists and public officials should be guides, should be means to lead someone to a better realization of his or herself, not simply a plastic end to be slavishly imitated.

I must discuss a pivotal turning point, a point in which America hinged, when America absorbed all of its inheritances, and a point in which all of America's inheritances bubbled over in the cauldron while the three witches stood brewing more. Greco-Roman and English government, Christianity, Humanism, Enlightenment ideas, the industrial revolution, they all fermented, all stewed. 1945 saw the end of the Second World War, the beginning of the Cold War, and the ushering in of the nuclear age and America as an undisputed world cultural leader. America now stood, with all

---

[57] In 1996, there was a partial lift of some sanctions on Iraqi oil. Iraqi oil was placed on the world market, and immediately western oil tycoons were hit economically. Soon after, Sadamms help was requested by a Kurdish faction to deal with another Kurdish faction. Sadaam bombed the Kurds (not all). The United States saw this convenient action as a way to impose the sanctions again, (and also bomb Iraq,) to protect the business interests of the oil tycoons.

[58] This situation is particularly true in China. Yet, the United States has a vested buiness interest in China. China pays almost 80% of the United States national debt (Diggins, Reagan).

her inheritances and contradictions, ready to lead the world into a new, global age. For almost two decades, America rode her post-war prosperity. The 1950s and early 60s saw what many hailed as the American golden age, ushered in by its overwhelming economic prosperity. Aside from a few scraggly, doped out beatnik poets, America was Pax Americana. What seems to distinguish this era is the abundance of confidence in America, in its government and her leaders, its military, in its superiority (at least by a WASP majority). American had her communist enemies, and it would vanquish them. President Eisenhower was a national hero, and the economy was soaring. This may seem to some a very broad generalization, but nonetheless, many historians would agree to its truth.

However, by the late 1960s, the post-war boom came to an end. John F. Kennedy was assassinated, his brother, and Martin Luther King. The South erupted over the civil rights movement, Hippies and new leftists challenged America. The Vietnam War dragged on, and dissent grew. The Soviets launched Sputnik and rattled American confidence. By the mid-seventies, a widespread dissatisfaction began to grow in America. The world market forced the US economy into submission, inflation soared, the dollar dropped. An energy crisis gripped the nation as OPEC nations held an embargo on oil. The US was brought to her knees, waiting every other day for gasoline, not able to keep warm in the winter. And in 1974, Richard Nixon resigned as president over the Watergate scandal (and his vice president resigned earlier over tax evasion). Bruce Shulman in his book "The Seventies," puts forth the argument that Americans lost confidence, indeed, as Jimmy Carter put it, there was a "crisis in confidence." Shulman tries to illustrate how many Americans lost faith many of their most valued institutions, in the military, the economy, in America's ability to produce its much needed energy, and overall, in the government itself. America was not as infallible as it has once seemed. America was humiliated, waiting on lines for rationed gas, unable to afford daily necessities, and shaking their head as their president resigned.

Shulman argues that during the seventies, after this massive loss of confidence in their government to help them, many Americans turned to the private sector. Americans turned to corporations, others sojourned on individual quests and "plugged in," trying to find individual salvation though various new ages movements. And I believe it is here were the Neo-mercantilist class finally completed its assent. As stated earlier, since the American Revolution, the neo-mercantilist class had a large influence on government and policies. After the Civil War, through the 1880s and 1890s, corporations began to exert considerable control over society. Many agreed that in 1889, JP Morgan was more powerful than the president, Benjamin Harrison. By the 1970s, and piggy backing off Shulmans ideas, I believe this "corporate revolution," took hold over America, and a major factor in it was the American population's willingness. Corporations "filled the void" left by America's hollow institutions. As corporations assumed control, many other Americans sought escape, many Americans, now disillusioned with America (and subconsciously with all of the failed inheritances and contradictions), no longer cared for America, or its government. Others, such as minorities, no longer sought integration, but rather diversity, such as the Black Panthers. America fragmented, and the neo-mercantilist class was free to pursue its hedonistic goals, now in the arena of global capitalism. The government became a tool of the corporations, thus my usage of the word mercantilist. However, it is dissimilar from the old mercantilism of the 16th century, because now it is business, and not government, that dictates policies.

And, as consequence of globalization, and America's position as world leader (even if somewhat tarnished) sent shock waves that reverberated around the world. At this time, the European and Asian economies were also influential. Today, in 2007, the neo-mercantilist governments around the world compete and cooperate with each other in the infant system of global capitalism. The effect of this global capitalist system spear headed by the neo-mercantilist governments around the world is one of a massive displacement of peoples globally. In third world countries, many times, neo-mercantilist governments seize the wealth of a country and it remains frozen at the top of society while the rest of society starves. Yet in all governmental systems there is a symbiotic, vertical relationship between classes. In many third world countries, this usually borders on patrimonial networks, and not true governmental structures. These networks see wealth trickle down various segments of society and it is these segments which enable the higher echelons of the neo-mercantilist class to stay in power. Joseph Mobutu, dictator of Zaire from 1965 to 1997, orchestrated a government based on this system. He and his cronies locked the wealth at the very top of society, it tricked down to various "strongmen," warlords, and army officers. These men than tricked down some wealth to lower strong men, until just a small fraction of wealth (not wealth, but bare necessities) reached the majority of the population.

Yet, in 3rd world countries, the vast majority of the populations are being rapidly displaced, mainly economically, but politically as well. These displaced segments of the population in extreme cases, take up arms and form militias to achieve their ends and secure basic materials for their existence. An example of this is in Somalia, and Sierra Leone. If looked at holistically, one can see as smaller more elite segments grow though global capitalism (Sierra Leone), or at least secure the means yet they need to control other parts of the population (warlords in Somalia) they displace more and more, less fortunate people. This phenomenon of "warlordism" had stymied many developing nations in the third world. However, not all displaced persons take up guns, this is an extreme case, rather, many times, displaced persons begin to look to local bosses and strongmen to provide them and their families with basic needs. Many times these local bosses provide the poorer segments of society with the things there government cannot give them (water, security, food) but in illegal, corrupt or extralegal ways. Thus the governmental system breaks down and degenerates into a system of decentralized feudal patchworks of small, rootless states.

Thus if global capitalism is taken to its logical extent, the world will see the emergence of this rootless, displaced class with no true government, heritage, values and in many cases, this class will turn militant to achieve their ends. This has been termed the "new barbarism" thesis. This rootless class will emerge throughout the world, in 1st 2nd and 3rd world countries, however it will take a variety of forms. The exponential growth of neo-mercantilist class due to global capitalism will see the growth of displaced persons in third world countries where 95% of the people cannot have their basic needs of food, shelter and safety met.

Is not terrorism in modern society a problem of displacement? I would venture that many who commit acts of terrorism have been either culturally, politically, socially or economically (sometimes a combination) displaced. The act of terror is a means to rebel against the society-this case a global society-that has alienated the displaced. Indeed, the Islamic revolutionary, Sayyid Qutb, after visiting America in the 1970s, feared that Christianity, Zionism and western modernity would annihilate-or displace-Islam. Is this fear not at the heart of terrorism? And in modern society this fear of displacement coupled with the problem of overpopulation may prove to be disastrous. As the world's population increases, there will be fewer opportunities for people-especially in 3rd world countries-to integrate into the global society. Rather, overpopulation combined with global capitalism will rapidly displace millions to the fringe of global society. These ever increasingly displaced persons may turn to terrorism, gang warfare, crime, drugs, sex or apathy to deal with their displacement and alienation (in 1st world countries as well as 3rd world countries). In a more esoteric sense, their alienation and displacement may make the ever encroaching threat of entropy more visible, while rendering the displaced more helpless in defending themselves against it. While many may turn to more positive means of dealing with this threat, I believe many will-and have-turned to the vices mentioned above.

However, the situation is markedly different in developed 1st world nations. The ascendance of the neo-mercantilist class in the global capitalist system will undoubtedly economically displace many in 1st world nations, but by and large, the vast majority of constituents will still live comfortable lives, or at least have the ability to obtain what they need. They will still have security and be protected by their governments. However, as the neo-mercantilist class grows, it will politically, culturally and socially displace more and more people. These people, unlike in 3rd world countries[59] will not form militant roaming bands or patrimonial networks for the simple fact that their material needs are met, and usually met in abundance. The neo-mercantilist class will provide for the citizen class as it alienates them. The citizen class will, semi-consciously, sink into a state of stasis, paralyzed by its own security.

Indeed, I think many have already succumbed to this static apathy. They are alienated from their government, alienated from their pursuit of eternal order and progress, even though they posses the means to achieve it. And this is how the world will end, in a lifeless, apathetic paralyzing stasis. The neo-mercantilist class will grow so fat that it paralyzes itself. The citizen class; content from their scrapes will do nothing and die, will achieve nothing, and leave nothing for their children to inherit but old, hollow values, dogmatic creeds, ancient hatreds and above all, apathy. To make sure the citizen class remains oblivious to their plight, the neo-mercantilist class pacifies it and divides it. The neo-mercantilist class educates the citizen class with dogmatic, rigid, industrial methods that narrows the scope of the citizens' comprehension and pigeonholes the citizen into discernable, predictable and exploitable patterns.

---

[59] This is not a slight to the populations in 3rd world countries. If Americans could not have their economic needs met, I believe we would behave in much the same way.

The Neo-mercantilist class divides the citizen class so as to assure a divided front. This maimed class is then incapable of realizing its own exploitation because it is too concerned with fighting itself. It cannot recognize that the root of its unhappiness and emptiness is due primarily to exploitation by the Neo-mercantilist class. The Neo-mercantilist class divides its constituents by race, class, gender, and status. However, I do not believe this to be a massive conspiracy "in the ivory tower" by the Neo-mercantilist class, but rather an evolving dominance by it, and evolving realizations of the weaknesses of the citizen classes and fringe classes.

The pacification of the citizen class has been evident since the creation of this country. An early example of this division occurred in 1676. Nathanial Bacon, a Virginia citizen, voiced his opposition to the colonist government because he felt, along with many other constituents, that the government, which was composed of elitist tobacco merchants, did not protect them from hostile Indians (the government actually wanted to keep good relations with local Indians because of the lucrative fur trade). So Bacon (who was a troublemaker and slaver overseer) rounded up a militia (composed of indentured servants and African-American slaves) to battle hostile Indians, as well as Governor Berkeley's colonial forces. Bacon and his followers were crushed. Yet, the planter class (forerunner of the Neo-mercantilist class) realized the great danger of this rebellion. Poor whites and poor blacks had united to resist them. If the much more populated poor class would unite, it could spell disaster for Neo-mercantilist class. And so, after Bacon's rebellion, the Neo-mercantilist class drove an irrevocable wedge through the poor. It divided them by the most obvious marker of difference-skin color. Soon after, African Americans were reduced to second class citizens[60]. The poorest white would always have his skin color. He would never unite with the blacks again because he scoffed at them, he hated them. They were "niggers" and less than human, they were animals. The Neo-mercantilist class fostered this hatred (and has done so) because it served a higher purpose[61]. They manipulated and exaggerated existing hatreds to further keep their more numerous constituents divided. And I believe this paradigm to have been a learned behavior by the Neo-mercantilist class, and I believe it to be applied to segments of society today. However, it is done not just with race, but with class, prestige, status, gender, and it is perpetrated by advertising, policies, public fears (such as the terror threat) and so on. Who dresses differently, who is Mexican, who has the new car, all of these meaningless attributes are fostered and manipulated.

As the small elite grows rich though global capitalism, part of the world will mobilize into factions and kill each other for land, water and race, while the other half will sit on their couch, languish in their possessions, hate their neighbors, turn the channel, and die slow, meaningless deaths. Order will become hollow rhetoric, just a stupid, worn out, hippie dream that the neo-mercantilist class will not even pretend to care about. All will die having contributed nothing but the same worthless things their parents did, all will die having contributed nothing other than their own temporary

---

[60] Treatment of African Americans prior to Bacon's rebellion was not ideal, and slavery did exist in limited forms, but after Bacon's rebellion, it became institutionalized, regimented and stifling.

[61] I do not contend that pure racism was not a factor. Whites hated blacks because they black, and different than them, and did not need the Neo-mercantilist class to tell them this, but rather, what I contend is that the Neo-mercantilist class realized the potential of this divide and exacerbated it for its own ends

preservation, and the fight will be lost, entropy will have used the intellect to conceive and orchestrate its final triumph, as the last inhabitants of the 3rd world stagger about their charred forests for food, the citizen-class in the 1st world countries will be too lazy to even care about surviving, the only thing to live for will be nothing because every inheritance proved to be a failure. Religion, government, science, reason all failed. The only successful institution proved to be industry and capitalism. The people, after all the contradictions became evident, unconsciously gave themselves up to it, or tuned out of society completely. Global capitalism and industry will drive the world to kill itself in fragments, like a schizophrenic cannibal hacking off his own flesh and eating a snack. And when there is nothing left to eat, entropy will have won, the people, neo-mercantilist, citizen and fringe, will have nothing, materially or culturally, and entropy will have won. Nothingness will have vanquished all.

Yet all is not lost. I am not preaching an apocalyptic doctrine. The government of the United States has the potential to truly liberate its citizens-or rather-it has the ability to enable citizens to liberate themselves from the ever looming threat of entropy and nothingness. And yet, it chooses to divide its constituents. The citizen and fringe classes are pitted against each other in endless civil wars, many not fought with force but rather with economics and subtle racism[62].

In the ranks of the citizen class and fringe class, this economic division among relative equals is evident when examining some of its relationships. One aspect of the population depends on the other, and vice a versa. I have called this the dependence theory. The Air conditioning repair men depends on the teacher to educate his child, while the teacher depends on the air conditioning repair man to keep his air conditioner running. Yet, when money is introduced into this relationship, it can become one of dependence. If my air conditioning malfunctions, then I must call the repair man, who charges me a large sum, which keeps me dependant on him, and which keeps me working harder in the symbiotic relationship to ensure I have enough money. If I am charged enough, the fragile balance of my checkbook-and my existence-will be in trouble, and I will have no means to my fight exploitation, only ensure that my family has enough money, gained from the symbiotic dealings with the

---

[62] I am not calling for the abolition of capitalism, while exalting communism. Rather, history has proved communism to be futile, at least extremely corruptible. All I am contending is that capitalism is not the end all; it is not the pinnacle that our society is striving for. It is not an end, but a means. Indeed, capitalism almost saw its own demise in the 1920s and 1930s with the onset of the Depression. Some turned to communism, some turned to fascism, while others turned to Islamism. In the United States, the depression shook people's confidence in capitalism. And the only way to save it was by implementing a degree of socialism into it, viz, the New Deal. Today, while espousing capitalism and reveling our defeat communism, many are ignorant to the fact that aspects of American government have been influenced by the ideologies of the welfare state. Social security, government assistance, and government housing are all aspects of a welfare state. While the United States is a capitalist country, it nonetheless is not truly capitalist.

However, I believe America's long derision of communism is more of a fear of itself. Communism, (and not Stalinism, Maoism, Socialism or even Castroism, indeed, true communism has never existed, rather only socialism that deteriorated into totalitarianism) but Communism in its purest form, calls for humanity to live like brothers, it condemns the hedonistic, unethical pursuit of wealth by the ones who control the means of production, it calls for an end to exploitation by the ruling classes, it calls for an end to the various conflicts that keep man divided. What is so wrong with these statements? I think America is afraid to peer inside her own heart and acknowledge that she may be guilty of exactly what Marx accuses capitalism of.

neo-mercantilist class. Yet, the air-conditioning repair man is in the same predicament as me, because his taxes pay my salary, and if his business falters, than theoretically, he will not be able to pay me. While this theory cannot be applied to all sections of society, it nonetheless holds true for a good portion of it. The theory of dependency is a gradual outgrowth of the division of labor and increased specialization. It is not a conscious exploitation, but rather an inevitability of the class system, which benefits the neo-mercantilist.

The citizens' class and fringe class are deprived of their voice and meaning, but instead given materials, security and minimum standard benchmarks and [63]quotas to aspire to (example: standardized tests) while always fighting each other. America is one of the greatest countries in the world because of the opportunity if gives to many and the security it offers to many. I am not calling for some violent overthrow. It has taken over 200 years to build this system and I do not think that work has been for nothing. However, I do think we have strayed far from the course. I would not want to be born anywhere else, but I have come to resent many occurrences in the United States. I resent the loss of my imagination. I have grown to resent the artificial constraints I am forced into. I resent the neo-mercantilists but am thankful for the opportunities they have given me. But I want my imagination back! They have made me useless and dull! They have robbed me of meaning, alienated me to an affordable home while they have built the greatest country in the world. I am socially, politically and culturally useless while I remain economically sound. My imagination has been dismantled while I am made to believe the infinite is just a stupid and worthless dream. By joining the social contract, I have given up some of my freedoms for the security the government can offer me. However, I never gave up my ability to think, my ability to imagine and create. Yet, this has been extracted from me ruthlessly by monotonous advertising, standardized tests, arbitrary authority figures who care only for bureaucracy (my current employer), insurance rates, dull professors, and ultimately society itself.

Throughout human history, man has sought methods to preserve himself in the face of entropy. He has erected governments; he has placed unquestioned faith in religion, reason and science. Yet all of these bulwarks, while producing many successes, ultimately failed or at least were exposed, they could aid, but could not lead the people to true or lasting order and progress. Progress faltered within them.

Each succeeding generation has left various institutions for the next generation to build, destroy or alter. Yet, in 2007, all of these inheritances-and contradictions they have produced-have surfaced, they have been subconsciously engrained within societies unconscious. And now, as with other time periods, society stands with these inheritances and contradictions in her pocket as she looks to the future. Yet with one crucial difference: society has the ability to annihilate itself.

The only inheritance that has produced some semblance of order-at least for the ruling elite-has been industry and capitalism, and thus has resulted in the creation of the present neo-mercantilist class, which is now global. Society is regimented and pigeonholed into its pre-fabricated and lifeless patterns. But its "success" is not true

---

[63] I do not imply there is a simple battle between citizen and fringe class, this is only one of many battles. Thousands of battles rage within each class, with various factions pitted against each other, fighting over race, prestige etc.

success, its success does not liberate people from entropy, it only pertains to small elites who then distribute their scraps. But scraps cannot and do not nourish. They are not enough to vanquish entropy, only stave it off temporarily. The small elite, their cronies, and open mouth constituents survive in a symbiotic and reciprocal relationship. But they all have nothing meaningful to contribute, nothing meaningful to do with their lives, nothing meaningful to pursue except their own, individual hedonistic quest for endless "stuff," their bad infinity. All over the world, in 1st world and 3rd world countries alike, billions are being displaced, billions with nothing, some with nothing materially, some with nothing culturally, politically and above all some with nothing imaginatively. As a result, some align into militias; some just lead lonely, meaningless lives. All have been stripped of meaning; all have been stripped of importance. All now have nothing because their inheritances have failed them. All the displaced-and even the neo-mercantilist-have only their own ability to create something new, better and infinite, Yet, this ability is not cultivated because it cannot be counted, it is left to wither, all wither along with it, rich and poor, neo-mercantilist and citizen, all are left to wither, but all do not have to.

And this is the crucial juncture. Society can drift on a path of material security, and die meaningless deaths everyday, or it can rise up and imagine a new existence. What has been achieved so far in 2007 is a type of bulwark against entropy, but it is a temporary one, incapable of protecting civilization for an extended period of time. Our houses, cars, bank accounts, insurance policies are necessities in our lives, they stave off the chaos, but ultimately all are swallowed up. Once again, I do not call for a violent overthrow; rather I call for renewed progress. I call out to the alienated citizen; I call out to him to take back his stolen imagination, to infuse the skeletal government with flesh, bone muscle and brains. Society has been given a foundation of materialism and security, now it must supply blood and infinity. The present bulwark must be saturated with the infinite. Only then can we truly progress. If we fail to capture the infinite, we will remain in a stagnant order, suspended and dangling above the entropy, idly floating, submerged in televisions, new cars and SAT scores. It is a progress all must build, neo-mercantilist, citizen and fringe. All have the capability to contribute, all must to survive.

Philosophical Definitions

While I have touched on many of these points in the paper, I wanted to clarify my position on the various philosophical terms that I felt relevant. Ontology is the concept of being. In my system, "being" is an outgrowth of the intellect. To conceive of "being" in an abstract sense is another defense against entropy. Our intellect (itself an evolution of the survival instinct) evolves further, at least in a theoretical sense, to encompass man in an abstract sense. I believe "being" to be only a concept of the intellect, but a nonetheless a crucial one, because it highlights man's relationship to reality. Man has evolved and fought to be the guardian of order, his "being" is a hard won conception that has been forcibly separated from the continuum of animal existence. The intellect, in its battle from entropy, has torn itself away from the animalistic conception and placed himself in a position to fight entropy with logic, judgment, reason, foresight and above all, imagination. It is within imagination that "being" is created and sustained.

Epistemology is the theory of knowledge. What can we actually know? Knowledge is accumulated information made possible only by memory and judgment. Knowledge is a defense against entropy. It allows the intellect to have an account of the past and thus make accurate judgments to aid in future assessments of situations. When man first used tools or used fire of language, when man began to manipulate his physical environment instead of merely reacting to it, he began accumulating and irreversible knowledge. It would be this knowledge that would make civilizations possible, which would make religion, government and science possible. Epistemology is the theory of knowledge, and knowledge is accumulated in the intellect as a defense mechanism against entropy.

Ethics are moral laws. When one looks at nature, there can be no way any moral laws are applicable. Instead the "law of the jungle," holds true, kill or be killed, namely, survival. Yet man, with his evolving intellect, concept of being, and religion, have created a system to adhere to. Ethics are a made man system of behavior, and they act as a bulwark against entropy. Whatever their source, government or religion, they are meant to encourage human beings to adhere to a higher law than survival, to consider their fellow human creatures and treat them accordingly. It has proved to be an amazingly successful bulwark because it calls for people to live in a communal atmosphere, it calls attention to the survival and well-being of other entities besides the individual, it calls for the order of an entire group or society.

Ethics are a subjective system, particular to place time and culture. Yet, while ethical rules are the intellects creation, ethical systems is also a universally applied concept. As I have stated before, what is universally established is the concept of living things progressing and maintaining order without the possibility of chaos and zero. What is subjectively established are specific rules pertaining to the universal concept of order. In different times, places ad cultures, different ethical concepts may need to be established in different ways, yet all are essentially aimed at upholding the universal standard of order. And so ethics are subjectively enforced, but are all used differently for the same goal, namely, as a bulwark against entropy.

Metaphysics is the study of what is real. Metaphysics is the most elusive term for any philosopher to apprehend. Since childhood, I have gazed at the world and simply wondered what it was, or if indeed what I was staring at was actually there. I wondered if this were all a dream, and if it were how I would know. And I do not believe I was alone in this endeavor, I believe many have wondered the same things. Indeed, one of the greatest philosophic minds, Rene Descartes, based his entire philosophy on just that concept, of whether the world was a dream. Sometimes the sheer awesomeness of a spring day will overtake me a like a tidal wave or fist to the jaw, science and logic fail miserably to gauge my feelings, I turn to poetry, but am still left in wonder. And it is here, I want to propose my conception and usage of reason. Reason, as has been shown earlier, is not the objective standard of reason it once was thought to be. Instead reason has been exposed, but I do not think it is useless.

Throughout the ages, there has always been a degree of uncertainty in assessing reality. Man raised his bulwarks against entropy. Thus, the need for institutions, guidelines, and creeds to keep these bulwarks working. As I described earlier, each of these bulwarks was exposed in various ways. However, I want to examine this again. During antiquity, many people in the western world were under the arm of Roman law. While there was various attempts to transcend reality, using philosophy, poetry and music, Roman law ruled for almost 1,000 years. During the Christian era, Christian dogma and creed replaced Roman law. What exactly was the role and purpose of the Roman law and Christian dogma? I believe that they were attempts by a civilization to bring order to existence in the face of uncertainty and in light of the inability to answer certain questions pertaining to reality. Questions such as: where did man come from? What happens when man dies? Why does man die? How was the universe created? How did man get to where he is? What are the laws of nature? Roman and Christian dogma gave order, told people when to stand and kneel, and told people how to think. A blueprint of existence was laid out for a population that was severely undereducated[64], and it worked. It kept the populace in line, it told them how to go to heaven and what to think about their life, especially how to deal with its sorrows and unknown and unfathomable variables. Many accepted this states because they did not know how to think.

Later on, as humanity became more educated, he began to test his dogmas, creeds and laws, thus, emerged the movements of the Renaissance, and the Enlightenment (also discussed earlier). In summation, the Enlightenment led to a slavish adherence to reason, despite the counter-enlightenment onslaught. And today, society has inherited this dependence on reason; in fact I contend that reason has supplanted religious dogma and totalitarian governments. Like water, it has filled the vacuum of unanswerable questions, it has given man a new method to combat entropy-but it is only temporary. This void created by uncertainty and unanswerable questions must be filled in order for humanity to have a working existence, or else, humanity would revert to an animalistic state. Today, even in the presence of democratic government[65] I believe western society relies on numbers, test scores, insurance rates and grade point averages. Humanity still depends on reason, and to do this, it uses rigid, easily quantifiable terms, thus the alienation of poetry, passion and imagination. It is a nothing less than a tragedy that humanity, liberated with education[66], still relies on a form of dogma, namely, numbers and test scores, to measure its existence. Humanity still craves easily discernable blueprints, now available in insurance policies, SAT scores, and placement tests. They fill the void left by unanswerable questions and uncertainty; they help to piece together an otherwise illogical world. These quantifiable terms rule like Hitlers, smashing their way into existence and usurping it.

And here I contend that humanity must at least alter this slavish dependence on reason, use it as a secondary means. What humanity must do is fill this void with imagination, with the infinite. In fact, humanities numbers are not as rigid and easily quantifiable as one would like to think. If numbers are followed to their "logical" end,

---

[64] Other civilizations, such as indigenous tribes in America and far eastern civilizations used totalitarian government and religion to answer many of these questions (discussed earlier).
[65] Many would disagree, but I think our government is hardly any more democratic than ancient Rome. I believe we live in an aristocracy-but it is one that the people of US desired. Our "democratic aristocracy" is the result of voter apathy and governmental alienation of its populace.
[66] See the discussion on industrial education.

they lead to infinity[67]. Humanity must not slavishly rely on test scores and grade point averages, these things, while being convenient markers, cannot truly measure existence. Indeed, numbers are much like Antoin Roquein at the foot of the chestnut tree.

Instead I think reason needs to be assimilated into a higher concept of reality, because in light of dogmas and reasons failures to be true bulwarks against order, straight dogmatic reason cannot be an end, only a means to order. Perhaps reason has to be seen as incomplete, but realized within the infinite. At this juncture in humanities present state, I believe in some cases we may have to suspend our reason, in order to arrive at a higher conclusion than one afforded to us by our reason[68]. Thus metaphysics is not dead, but has been forgotten because it does not integrate with dogma and reason.

Philosophers from Plato to Sartre have questioned the validity of their perceptions of the external world, and not one philosopher has thus devised a satisfactory answer to the question of reality, how could they? Truly, when I stare at the world, how do I know what is real? My philosophy works under the assumption that reality is indeed real. I take a Kantian approach to metaphysics, in respect to his doctrine of practical faith. We must believe our world is real, because if we did not, then why struggle to produce any lasting achievements? Why would we so ardently pursue order? If I do not eat, I will die. If I jump off a building, I will die. Why should I waste my time wondering if these events are real or if they are illusions? There are millions of people around the world, whether if they are in war torn countries, whether if they are beggars, cripples or handicaps, whether if their families have been slaughtered by rebel armies, which do not have the luxury of worrying about whether this world is real or an illusion. They do not have the luxury to simply think all of their pain is an illusion. They are starving and sick. This world is real to them and it is real to me. For thousands of years, various religious doctrines have taught us that "this world" is worthless, finite, revolting etcetera, and simply to worry about the next world. I believe this to be a major difficulty in our progress because "this world" with all its shortcomings, contradictions, violence and crime is all we have. We must stand up to it, not cower in pretend fortresses. What is real is what hurts, what is real is order in the face of chaos and nothingness.

Nominalism is the last term I think relevant to this discussion. Nominalism is the idea that the names we have given to certain entities (i.e. Justice, truth, love) are simply that, names. This doctrine was expounded on by the 12th century philosophy, the bishop Thomas of Ockham. I think it is relevant because many times we talk of love or justice as if it is a material thing, but really, these are made up terms for things we have created. Our creations are personified when in reality they are creations to help stem the tide of disorder. I do not think it be of a major concern how they are conceived, but nonetheless, our conception of love, justice, truth beauty etc, is similar to our creation of ethical systems. While the human intellect has created these entities

---

[67] This is discussed in more detail in appendix b.

[68] I realize the implications of asking people to suspend reason. This could be exploited and abused, and perhaps could lead back to a fanatical dogmatism. While my arguments are weak, all I propose is that reason is incomplete and has been exposed, but is nonetheless useful for our purposes.

and named them, they are human creations, particular to circumstances, but universally, they all adhere to preserving order in the face of entropy.

My next task is the rebuttal of anticipated criticisms. I believe that my biggest critics will come from religious adherents. I believe many religious adherents will accuse me of eliminating God from the world, and turning the world into a completely mechanistic existence, with only human effort being the determining factor of reality. While I do place the highest emphasis on human will, imagination and work ethic, I do not believe that I have completely removed god from the picture. I believe the existence and reality began when nothingness tried to expunge the possibility of order. Order, at the least the possibility of it, has always existed as the antithesis of nothingness, thus, creating chaos, or incomplete chaos. Order, even possible order, is God, because it is absolute perfection, it is an absolute vanquishing of nothingness. Over the years, ritual and dogma have weighed too heavy on the mind, forced the mind to forget the real omnipotence of which it is capable of working toward. As mentioned earlier, Marx admonished Plato and Christianity for placing divinity outside of humanity. I think the idea of a transcendent, omnipotent God is the infinite. Jesus, in his struggles against the Pharisees, chastised them for their adherence to hollow dogmas instead of focusing on God. To use religion against religious adherents, I think Jesus' warning needs to be repeated today. Too many people today adhere to hollow, empty dogmas and rituals because they are familiar; they give comfort to an easily assuaged mind. But performing some mechanistic action will not realize the infinite that every human mind is capable of achieving. Soren Kierkegaard's philosophy is similar to this as well. He wanted to destroy all dogmas and rituals because he recognized those were not vehicles to the divine, but merely to our own comfort and familiarity. He called for humanity to carve out their own salvation, with fear and trembling.

I have not destroyed God; I have not left humanity in a cold mechanistic world. All I have done is taken away the slavish reliance on some higher power. If humanity wants a God, then let it actualize him! The possibility is there, let us fight for it! For my argument, I rely heavily on Anselm and Kant. If one can conceive of a higher state of order, then it is attainable through the infinite. God (or Allah or Bramha) exist as the possibility of order. God's omnipotence is of no matter to me, it does not fry my eggs in the morning. I believe what I know to be true. Call me a doubting Thomas, but faith will not stop a bullet from killing me. I have faith in order, and in humanity's ability to reach it. I do not deny the presence of God, for "he" may be waiting in the infinite, but I have to realize him first. Perhaps it is a test. To quote Nietzsche, why would the all powerful omnipotent creator not want to be in the company of creators? Would the creator of the universe want to be surrounded by dogmatic and brain dead dummies? Most likely not.

However, I think the biggest criticism from the religious camp will be my treatment of divine punishment and death. Death is reabsorption back into the chaos which spawned all. The human body dies, but the idea stays alive. The idea of the body, or more accurately, the idea and memory of what a particular body has done, is what survives, this is akin to the soul. The memory becomes a symbol, for all to adhere to. The symbol is revered or reviled by later generations. The symbol becomes ethereal, it becomes embedded in present and living minds, and it becomes part of the framework of reality[69]. It is this memory of a particular body; it is this symbol, which will be resurrected in the infinite. While it is not up to me to be a final or omnipotent judge, rather, I believe any person's legacy will become evident over time. And to clarify the term "legacy,' I mean their contribution to either order or chaos. Now the crux of the matter is this-the infinite is being sought after by both order and chaos. The idea of Adolf Hitler resurrected in an infinite controlled by chaos, with no boundaries, is a frightful idea. I believe humanity internalizes these symbols and they then become part of reality, and are either exonerated or condemned.

Conclusion

    I believe the world to have started as the expungment of order. When chaos tried to expunge even the possibility of order, then the world, existence and reality began. The battle between order and chaos began. Each force tried to vanquish the other. Order, now cast apart from chaos, strove for a world with out the possibility of chaos. And chaos, which was an incomplete form of nothing, strove to vanquish the possibility of order. In its first incarnation, order emerged merely as self-survival, unable to better itself, only able to survive. But important precedents were nonetheless set, such as evolution. Order knew it was not strong enough in its finite form to defeat chaos, and so it evolved into stronger and stronger forms until the human being emerged, and what separates the human from every other animal is his intellect. The intellect was not only able to perceive, but to reason, to judge, to anticipate, to use foresight and most importantly, to imagine. It was with this intellect that order erected civilizations, governments and religion as bulwarks against entropy. With the emergence of the consciences mind, human beings were now also able to manifest their belief systems. Certain strong minds manifested a system of order, or sanity for all to adhere to. While this system is necessary to ensure the survival of humans, it is nonetheless flawed, and at times, in need of revision.

    However, it was also with this intellect that order conceived of a release from the threatening chaos, order first envisioned the infinite. Yet, chaos also needed the infinite to conceive of a world without order. And so now the battle became pitched. And to quote Caesar "the die had been cast." Order and chaos battled with governments, religion, science reason, nationalism and industry. All were manipulated by chaos, all have failed to preserve order, but all are invaluable in the struggle for it. 2007 AD is a crucial time. All of the various bulwarks against entropy have ripened and bore fruit, all have bore their contradictions. Humanity was decided if it will slide into an abyss of apathy, nihilism, and contradiction, or if it will realize the infinite within itself. If humanity will heal its contradictions, access its mistakes, failures and shortcomings, and forge ahead. Humanity must realize this crucial juncture, it must look beyond its pettiness, look beyond its imposed sanity and realize the infinite and manifest that to later generations, instead of its contradictions.

    I am not under the illusion that many people will be affected by my thoughts. In fact, I am rather pessimistic, and believe that my ideas will probably never be published for distribution. Yet, to echo Schopenhauer, I do not seek fame and riches

---

[69] I will lead a further discussion on the "framework of reality" in appendix A.

with my work. Instead I seek to just influence people, or at least one person. My dream is for my ideas to transcend time, sanity, and skin, and to positively affect one person, and make that person think about the world slightly differently than he did before reading my works.

A Short Description of the Appendices

Appendix A is an in-depth examination of reality. I have broken down reality into mathematical equations. I try to understand what the "present moment" actually is, in all of its fleeting beauty.

Appendix B is a further in-depth discussion my conception of the infinite.

Appendix C a refutation of logic, which, as I have read more, realized was not original. Nonetheless I believe I make an interesting point, or at least I interestingly reiterate an already made point.

Appendix D is my theory of mental illness, specifically Obsessive-Compulsive disorder, and how these illnesses mesh with my philosophy.

Appendix E is actually a short story that I had written almost four years ago, so it was written previous to the formulation of my ideas, but nonetheless, I think I do raise some interesting questions about the nature of faith.

Appendix F is a short story. I have aptly titled it "parable of future generations." It is a simple allegory that deals with the divorce and tension between the sciences and the humanities, and the danger in "burying" the past. History is something to be learned from, not forgotten.

Appendix G is a letter written to my unborn son. I anticipate Herder and Goethe in this. Herder believed that literature was a greater tool than science in apprehending nature. Goethe believed in a unity of science and poetry. And so my letter is mostly nonsense, but a nonsense I think we cannot afford to ignore.

Appendix H is a list of acknowledgments

Appendix A

<u>Discourse On Reality</u>

I
What is reality?

Reality is an infinite amount of interconnected events stitched together to form a present like an unmoving tapestry. The human intellect chooses a course and carries it out, but there are an infinite, or near infinite amount of courses that could have been chosen, yet they exist only in theory, not in actuality. And so the present exists in one motionless second and can be represented by the equation $1/\%$. And it is this present moment which drives reality forward, which humans have erected-positive or negative- to stand against entropy. This present second is the creation of the human intellect acting against nature, namely entropy. Each second that is experienced is an advanced bulwark against disorder. And in each moment disorder is trying to reclaim order, trying to rape it into nothingness. Yet the progression of time and reality must be dissected and understood. Each singular moment is fortress, a man made enclave composed of manifested sanity, powered by the competing forces of the human intellect and inevitable decay of chaos.

1 stands for the present course of events or combinations that actually occurred. Infinity stands for the infinite amount of combinations that could have occurred. To illustrate this concept, a simple example can be used. If a person wakes up in the morning at eight a.m., then takes a shower, then brushes their teeth, this is one combination. If that same person woke up at three minuets after eight, took a shower and ate breakfast, this is another combination. If that same person doesn't wake up at all, this is another combination. So on and so forth. Reality is merely one combination out of infinite combinations.

II
Who Determines Reality?

We can view the future as $\%/\%$. Infinite events out of infinite events are possible. Reality or $1/\%$ is merely the sorting of these events or combinations. So, in theory or speculation, an infinite amount of past realties can exist, with the deduction of the actual reality that actually existed. So these infinite, unlived combinations can be represented by the equation $\%/1$.

But who decides this course of events? Who decides what 1 will be in the $1/\%$?

What determines reality? What determines reality has to be something instinctively involved in it. Reality cannot be an already laid out plan, the tapestry cannot be know to the artisans. Instead reality must be determined by its consciousness participants, namely humans. However humans are merely the present culmination of order. Yet, the human intellect is the best bulwark against entropy. Human reality is decided by human activity, by conscious thought, and ultimately by the imagination, however, it is buttressed by the encroaching swell of chaos. This dynamic determines reality, at least for a moment. Humans actively choose the 1 and carry it out.

III
How is reality determined?

The participants in reality actively choose their own path. They actively and knowingly make their own decisions (whether by forces in themselves or outside their control). In the end, a human being is responsible for their actions by the law and others.

So the question remains, how is reality determined? How do these participants determine their 1/%? Do they simply choose randomly and at will? Or is there some type of system that gives them guidance and a point of reference?

Reality can be seen as a mathematical equation, but it exists in a visible form as well. Examples of this form are trees, automobiles, dirt, steel, stars, cats, cell phones and human beings themselves. These things form and "already existing framework". And this already existing framework exists exclusively in the present moment, in the 1/%. Some pieces of reality have been human made and others cosmically, naturally whatever. In any 1/%, every single piece of reality is an already existing framework. And this framework, aside from being our surroundings, is also our fuel. All of the pieces of reality interact with each other. Example: automobiles drive on roads, metal, wood and vinyl are assembled into houses. The human mind dictates and establishes these connections. Limitations, awards and measurements are made based on the interaction of this connectedness.

These interactions of pieces can be summed up as the process of sanity. Human beings value certain actions, despise others, and still expect others. Example: if your shoelace is untied you are expected to tie it. If there is a stain on your face you are expected to use a napkin. Also, the establishment of time has been developed. You have to be on time, certain things are only done in certain seasons. "Time" as we know it with minuets seconds and dates does not exist in nature.

The relatedness of these connections is what human beings deem as sanity. Therefore sanity is the rigorously pursued manifestation of norms based upon the connection of pieces within the already existing framework called reality or 1/%. Example: over years humans have deemed it sane to want a family, to want a high salary, to earn awards and recognition, when in actuality, these events are merely manifested and sometimes violently imposed ideas. They are an evolution of process and dependency. For what would we be without sanity? The earth was conceived in chaos. The greatest invention a human mind has ever built was the categorization of this chaos. But this chaos is the natural state of the world, of every human being. An infant can be deemed insane because it has no idea of the manifestation it was born into. Only after years or training and process will most infants develop into sane, rational participants that help reality function like an engine. The human becomes like a piston or some moving piece forever pushing it forward.

If the participants are likened to engine parts, then the manifestations act as a sort of "fuel" to power reality or 1/% into the present moment. The infinite possibilities and things they contain work within the engine of reality to drive it and create the present to be perceived and categorized by our human eyes and minds.

IV
A further definition of the fuels in reality.

These fuels of reality exist as our surroundings and ideas. They exist as streams of living and non-living things constantly around us. Example: a billboard advertising a weight loss product depicts scantily clad women as sex objects. A young girl sees this image and subliminally wants to emulate the image. The actual billboard and the idea that the billboard contained act as fuels, driving women to be like something else.

And this sanity or fuel can be used to manipulate, control, exploit and ostracize people. For example, if a man is fueled by one idea, and another fuels the majority of people, that one man can be labeled as different, or in the worst case scenario, insane. But does this charge of "insanity" really exist? When sanity is merely the manifestation of a set of created norms based on the majority's perception of 1/%, it can hardly be deemed worthy. Sanity can twist and change like water to accommodate the 1/%. Thus, while sanity is crucial and necessary invention to achieving order, it must be remembered that it is an invention, and that it can be manipulated and exploited, in fact sometimes so much to actually regress from order (i.e. The third Reich in Nazi Germany). Sanity cannot be the last word, because sometimes sanity leads to a dead end which is incapable of achieving the infinite.

For example, two men can both go to college and graduate. The first man receives a 4.0 G.P.A and is awarded and recognized. Society labels him as brilliant and he acts accordingly to his label. The second man receives a 2.3 and society recognizes him as mediocre and not too intelligent. But the question is what if that man who received the 2.3 went to school because he merely wanted to subscribe to the idea of sanity? What if the man with the 4.0 was nothing more than a sponge and rotely remembered everything that was acceptable and exclusive to the present 1/%? The only reason society recognizes this man as brilliant or genius is merely because he excelled in the sanity manifested. A sanity that he has practiced studied and loved. He opened his mouth like a nozzle and burned the "sanity fuel" poured down his throat by society. In fact he craved it because he could not understand or believe in anything else. While the man who received the 2.3 only adapted to this sanity, just enough to get by, just enough to not be thrown into jail or labeled "insane". The idea, or fuel of high grades or sex or whatever powers most people, satisfies them, but it could not power the second man. What if he is fueled by something different? What if he cannot ascribe to the sanity imposed on him, to the awards and numbers? Instead, what if he ascribes to the infinite, but since this is an unfamiliar concept, is not understood by the majority, and thus shunned.

Society will exalt one man and not acknowledge the other. However, while society's back is turned, too busy heaping awards on the first man, the other man, banished into obscurity will tap into a different fuel line, nourish his rusted out stomach, rise up and obliterate the so deeply entrenched sanity that had him chained for years. He will murder the sanest visions that he ever knew.

V
Are these fuels free to all?

The rivers of fuel that drive reality are known to us as inconsequential pieces that collect dust in our living rooms and garages. The dirty tarp thrown in the corner, a dead spider, a shard of glass. All of these seemingly irrelevant things coupled with deeply rooted patterns (using a napkin, shaving, going to school) have the power to create the 1/%. All of these supposedly meaningless things are fuel, driving reality ever forward, pumping hearts and pistons into a finite present moment.

These fuels are present and exist and power the majority of people, make their hearts beat and cars drive but are they free? Can one person usurp them and use them? Do we have to compete for these fuels?

Everything is competition. Reality is a battle for awards, recognition and friendship. It is a battle for money and cars, a battle for fuel. Some people can channel (usually unknowing) more fuel; they can usurp reality and burn it. The "winners" can take more of the offered fuel and beat the empty ones into submission. The winners compete for existence, for the recognition they crave and have been taught to crave. They compete for reality in the manifestation of the sanity, within the interconnectedness.

And it is this inter-competition which is so detrimental to order. Instead of working toward the infinite, humans turn on each other, they suck the fuels from each other, trying to starve their fellow man. But it does not have to be like this. While I do not labor under the false pretense of some idealistic communist society, I do believe humanity could achieve more cohesion through education. I expounded on this argument in the body of my paper.

Sometimes it may appear as if existence is a story, sometimes patterns may appear (example of this, high school students go to college, high grass is mowed, dirty clothes are washed), however, under these supposedly chronological actions is really a competition for them as fuel. The better acclimated high school or college student, the one that can usurp more fuel in the form of test grades and awards, new cars or sex will be the winners, and they will win the battle of reality. They will leave the losers a withered shell, craving, devoid of fuel.

VI
What if a person cannot win these fuels?

Some people are left aside, left empty and malnourished because they cannot win these fuels. Society uses labels like idiot, freak, or bum to label these empty people. These people could not earn the privilege of reality, they cannot merit their existence, and they are not strong enough. Instead they are beaten. Driven to subjection, loneliness and death. Starved for fuel, starved for existence. These are the children who cannot get a date to the prom, who cannot score touchdowns or get A's. Industry society discards these non-producers. Yet, they are not non-producers, they just do not produce what society wants. Instead, society vests its interests in failures and unproductive members because of some false presumption or blatant favoritism or nepotism. This act severely hampers the progression of order and must be remedied.

VII
What if a person does not want to win these fuels?
What if they desire something more?

Not everyone in the already existing framework of reality will compete for these fuels. Some participants will reject them flat out. Usually unknowingly, sometimes the things he has been fed since birth can no longer nourish a person. The traditional fuels of grease and gasoline, semen and new cars no longer burn for the individual.
Like filling an automobiles gas tank with sugar, these old fuels become useless. In some cases, they are just enough to make the person appear "sane", just enough to get by, by still, the person is starved and empty.
And so, a small minority develops. An unknowing army of rejects with rusted out stomachs. Misfits and outcasts that have given up the traditional fuels in life (whether knowingly, unknowingly, voluntarily or involuntary) and now seek something more, like a cosmic octane. They do not know what they seek, but they crave it. They usually do not know that they even seek anything at all, but they do. And sometimes will go "insane" (by societal standards) or die in the process, starved and hungry, but still better off than the ones who fill themselves with what everyone else uses.

VIII
Does a place exist without sanity?

Participants who find themselves unable to use fuels of existence, wander through reality, stagger though concrete and skin, wash with it's soap, sputter and can barely be driven to the present. Where exists a neutral place?

First, it must be a place in which reality is simply reality. A place where participants are no longer participants but *human*. They can sit and watch reality surround them, love other people and not use them for fuel, a place where knowledge is not transferred into grades and power, and a place where semen and breasts and G.P.A.'s are not burned to produce a present.

In this reality, 1/% would be false and useless because instead of infinite combinations of connectedness, there would only be 1, a perfect expression of true reality, when knowledge is sought to better humans, not to rank participants. There would be no fuels to burn and exhaust, no one to usurp existence and win it like a medal or severed head. The hollowed out stomachs and rusted hearts would never be empty because emptiness would not exist, only solid bodies, living and non-living, represented as one immovable force, it would not power reality because it would be too busy *being* reality. In the 1/% reality is wasted because once used as fuel, it is stripped, burned and useless. Perhaps this is the idealistic infinite to which all of order tries to progress to.

IX
Can humans achieve this reality?

A certain unfortunate dynamic has been created. Human participants, even those who do not desire this "reality fuel" are instinctively and irreversibly attached to it. For the duration of life (which is nothing more than acting in accordance with the norms dictated by connection exclusive to any 1/%) resistive human participants are saturated with these unwanted fuels, soaked until they stink like gasoline. And even though they want to nourish themselves with something else, even though they so desperately want to reach the perfect reality of 1, they remain trapped in 1/%. They become unwilling participants forced to compete for fuels that they never wanted, because the struggle is familiar, especially in 2007, when the contradictions of existence had overlapped, many refuse to look beyond themselves, instead, they rather compete.

XI
What would the world be like without sanity? Can sanity be defined as a number?

Sanity was shown as the manifestation of a process. It became a rigorously pursued idea, imposed by a majority of people. Simply said, the majority of people in the world are sane so the world is sane. It is divided into rational terms. Over millennia, a general evolution of this sanity has gradually replaced the natural chaos from which we formed, and the specific developments of sanity acclimate to different

1/% exclusively. Example: over millennia, the idea of keeping warm, eating and being safe have taken root from the chaos that the world formed from, and the general ideas have manifested and evolved. However, the modes in which we keep warm, the things we eat and how we protect ourselves are specific developments and are exclusive to different present moment or 1/%.

Basically the same general fuels usually exist for everybody. Basic needs like shelter and warmth, food and safety but there are different versions of these fuels. However, the acceptable sanity (with obvious variations) can be represented with an equation. Example, there are a limited number of acceptable shelters for a person to use, whether it is a house, trailer or hut. The variable x can represent the limited amount of sanities. And the manifested sanity dictates which ones are acceptable. The ones that are acceptable can be represented as 1 main idea (able to be divided into different parts because there are usually more than one acceptable form) Thus, sanity can be represented as 1/x. It is one acceptable pattern out of a variable of accepted and unaccepted patterns. Example: if you buy a plot of land in a suburban neighborhood, you would be deemed strange or even insane if erected a tee-pee or trailer amid the other beautiful houses, all though you could do this. Most everybody will choose the 1, or rational path dictated by sanity. It is sane to build a type of house (What house you build is your choice, thus the 1 can be divided but it is still one main idea). The 1 is chosen from how many other unacceptable options there were. (Most people chose to build a house even though they had the choice to build a tee-pee, hut, lean to or trailer.)

Another example of manifested sanity is the idea of currency. We are told to believe that a dollar bill is worth a fixed amount. However, this is merely a piece of paper that has evolved into a desired currency. Notions and institutions have been built around a worthless piece of paper, simply because the majority of people, the sane democracy manifested its worth to the world. The idea of it's worth is manifested, evolved and accepted. One Greek philosopher said "Trash is more valuable to donkeys than gold". This statement is relevant because we affix the value of paper currency and manifested it and use it as fuel when really it is a human invention, not intrinsic to the natural state of chaos before the world.

When we accelerate at green lights, sanity tells us that the other drivers will stop at the red. The manifested ideas of right a ways, stopping and yielding make for a usually safe ride. We, as drivers adhere to the 1/x because we only perform the actions dictated by the sanity we live in.

What if a driver decided to accelerate at a red light? What if the driver went in reverse on the highway or slammed their head on the steering wheels? These are all actions represented by the variable x in the 1/x. However, we abstain from performing them because they are not sane.

Yet, what does make a driver stop at red light? Or not stab himself with a pen? What makes a human being participating in reality able to wipe his mouth, not slit his own throat shaving or tie his shoe? The manifested sanity or 1/x that has been rigorously drilled into the head of participants gives them a point of reference.

XII
What if a person does not adhere to the 1/x?

Some people, like the ones mentioned earlier, cannot be "fueled" by the normal and acceptable fuels of sanity. Therefore they are deemed insane. They cannot adhere to the 1/x. They cannot accept the sanity or 1 manifested by the majority of people. This minority of people, or the insane, can be represented by x/1. The denominator representing 1 is the 1 manifested in the 1/x that these people have rejected. The numerator is a variable because this represents the other various fuels that these people adhere to, or at least find more acceptable than the manifested ones. Thus, sanity is the inverse of insanity. In the world sanity or 1/x has been deemed acceptable and therefore adapted and manifested by the majority. X/1 or insanity is deemed abhorrent and forgotten.

XIII
How relevant is the idea of humans not adhering to the 1/x?
Aren't humans instinctively drawn to self-preservation?

Humans- and animals are instinctively drawn to self-preservation. Humans tend to avoid situations that endanger themselves. However, this is only done in extreme cases. If humans practiced absolute self-preservation, they would never leave their homes. They would lock themselves in a fortified room with sufficient food.

However, we know humans do participate in actions that potentially threaten their lives. Driving in cars, flying in planes, choices of occupation (i.e. firefighters and military) eating unhealthy foods or living unhealthy lifestyles. Humans' sacrifice varying degrees of self-preservation for honor, ethics, or pleasure (saving someone's life, having unprotected sex, playing professional football). Ultimately, humans sacrifice their self-preservation for the passage of reality, for the attainment of fuels and the 1/%. The 1/x is a manifested reference point for them to do this. The forsaking of self-preservation emerged as a result of the intellect, which could judge, reason and imagine.

Self-preservation is practiced only superficially. Humans preserve themselves only in extreme circumstances (and not even all the time; example-fire fighters running into a burning building). The question remains, what if humans-or participants-do not adhere to the 1/x? Since the need for self-preservation is shown to be fallible and corruptible and not absolute for various reasons, what keeps a participant from jerking his car into a guardrail? Or not smashing himself with a brick?

If a human participant is left alone in a room with nothing but a bed and knife, why doesn't he stab himself, or strangle himself with the bed sheet, or bludgeon his head against the wall? Instead, he simply falls asleep. Why? Simply because he adheres to the 1/x. He knows the 1 by heart, has been trained to love his body and avoid pain. He knows the process of logic and rational that keeps the knife out of his wrist. There is no good reason to put it in. He is trained to love the 1 in the 1/x and loves it with all his heart.

XIV

So the manifested 1/x keeps the participants functioning, keeps the knives out of their wrists, keeps them from willingly driving into telephone poles, makes them wipe their mouths with napkins, It acts as fuel to drive them ever forward to the next 1/%, into the present.

There are also other factors that help configure and manifest the 1/x. The major factor is the law. The law is a set of imposed rules decreed by the majority as a necessity. The reason why a normal or sane human does not punch a co-worker in the face is because of revulsion of breaching the 1/x and for fear of penalty of law. Legal laws are devised, learned and manifested by the majority, kept functioning by the majority.

However, just like the idea of self-preservation, legality only exists in limited forms. The institution of laws is not what they appear to be. As stated before, the law is a set of imposed rules decreed by a majority for deem able necessity. But the laws do not bring us up to a higher level, or offer us anything better than what we already have. The laws and the legal systems only offer us a negative peace, they only give us an absence of chaos with no definite alternative. They are the precursor to sanity. By creating an absence of chaos or 0, the 1/x can thus be manifested over this new "clean slate".

So out of x and % of the past and unaccepted sanities, first it must be reduced to 0. This negative peace can be achieved though imposed laws of the majority. Laws against killing, stealing, raping etc. Once this negative peace or absence of chaos or 0 is achieved, than the sane democracy can manifest 1/x, or the one acceptable pattern or sanity exclusive to the 1/%.

So the construction or equation of sanity can be represented as such:

$(\%+x) -(\%+x)=0$

Infinite combinations and unacceptable sanities exist but though a set of imposed legality, are reduced to 0. (people who breach the 0 or "flirt with insanity" are imprisoned and punished)

$(0+1/x)\rightarrow 1/\%$

Using the 0 as a foundation the "sane" majority impose their "sanity" for (for is represented with ->) any exclusive reality (1/%)

In conclusion, manifested legality and sanity are the configurations of the reality around us. Starting with the law, which establishes the 0, then building upon it with sanity, humans have constructed a reality from chaos. These manifestations instruct us not to kill, not to rape, then to tie our shoes and zipper our coats, then to go to school get good grades and love our cars.

XV
The difference between simple and complex sanity

Sanity was seen to be represented as 1/x or the acceptable pattern from the many. However this needs to be clarified and separated into two categories, simple sanity and complex sanity.

Simple sanity is the everyday, inconsequential actions in life that most participants fail to notice (such as shaving, pumping gas, ordering food) but none the less act as fuel. Those simple sanities can be represented by the 1 1/x because 1.1 will still be one.

Complex sanity on the other hand is the next degree; fuels of going to college, weight loss, making money, basically things that people strive for. This can be represented as 1 2/x.

Both of these sanities combine and can be represented by the 1/x.

XVI
What does the 1/x leave us with? And the dual nature of the fuels.

Through the construction of reality, though the 0 and 1/x exclusive to 1/%, we are left with a perceptible reality, able to be seen, heard, touched, tasted and smelled. We are left with a template of order against the encroaching chaos.

But what does this construction truly leave us with? What are the fuels in front of us? First, they must be clarified and divided. We divide reality or the bi-product of 1/x and 1/% into four main parts: Perceptible, Perceptible to thought only, living and non-living. (Example: a tree is living and perceptible, the idea of love is perceptible to thought only, a human being is living and perceptible).

These "fuels" as stated before are the already existing framework. They are also twofold. They act as the fuel or power of reality and are also the bi-product of reality, like the exhaust of an engine (example, the idea of going to college and getting good grades fuels some participants but this idea is also a manifested end product of sanity).

These opposing cyclic patterns drive reality without destroying it; they keep it stationary but functioning. What I propose is for the infinite to replace this static reality that has emerged in 2007, that has been built a top a throne of contradictions. This sickly equilibrium only keeps us chained into place, we must forsake the stasis and realize the infinite.

If a participant looks out of his window, he can glimpse this reality functioning. He can see the fuel bi-products of 1/x. helping to construct and being burned as the same time.

But what exactly does he see? What are those fuels composed of? The inverse principle can be used to decipher this. What he sees is what sees and nothing else. Example: If he looks at a red car, it is a red car and not a white polar bear or notebook. The reality he sees can be reduced into one simple term. The inverse principle can reduce the fuels to what they actually are. Through this principle represented as y=z+ because y cannot equal z or anything beyond it. So the inverse principle can be applied to the manifested sanity and the simulations burning fuels to reveal their true natures.

XVII
What is the purpose of the inverse principle and how can it be used?

The purpose of the inverse principle ($y=z+$) is to sort through the % and x, to cement the 1/x and 1/% into an immovable foundation of reality. It can be used to decipher the vast amount of unlived and unacceptable information by using simple logic. My hands are not my feet; the pavement is not my hands. I am a human and not the pavement.

It also gives us a deductive proof the perceived reality and an implied point of reference, acting as a sort of funnel to give us the 1 in the 1/x and 1/%.

XVIII
When can the inverse principle prove false?

The inverse principle can prove false when the x and the % have not been properly sorted. However this condition cannot exist because reality is always reality (1/%). The inverse principle simply states what could never be and assures us of the absolute truth.

XIV
Is there a danger in disregarding the % and the x? Is there a danger in forgetting the unlived and unacceptable if the inverse principle assures of the absolute truth?

By using the inverse principle to achieve a singular point in reality, participants unknowingly disregard the x and the %. We never stop to think about why we don't jam forks into our hands. Or why drivers halt at red lights. Instead we just carelessly proceed through reality, ignoring the things we could have become.

Ignoring the unlived and unacceptable is forsaking the things that place us right here into the present, in this very singular point, like the point on a straight-line segment. The disregarded which we call the Void (or V) Factor sorts out and steers participants (through their own volition) to the 1/% and the 1/x. The danger in forsaking the V-factor is to never realize the corpse you could have become, never to see the drooling, stumbling beast that repeatedly stabs himself with a butter knife. The V-factor is everything that we are not but could have been. Ignoring the V-factor is taking sanity for granted. Sanity, reality and order are privileges that need to be won and fought for; they are a one in a billion chance that might never have happened. Ignoring the X and the % is ignoring the blood from the fight.

XV
Are some of the participants better suited to perceive the V-Factor?

A small minority of participants are either unwillingly or willingly able to perceive this V-factor. These are the people who adhere to the x/1, who need different fuels to power them. The ordinary fuels can no longer drive them. They do not love the 1 in the 1/x. Instead they explore the x. These people, who society labels as "freak", "idiot" or in the worst case "insane".

This small minority is always kept submissive and subordinate to the manifestations of the majority. They are unwillingly infected with the virus of sanity and if they reject it are deemed mad. And so the sane majority manifested their sanity, burn their fuels for everyone willing and non-willing.

XVI
What would happen if the 1/x were no longer manifested? What would happen if the small minority, instead of being submissive, destroyed the fragile 1/x and manifested a different reality?

If the ones deemed "insane" or "outcast" were to somehow revolt and manifested their own version of sanity, it would destroy the current 1/x with x/1, which would then be adopted as the new 1/x. This small minority (not just institutionalized people, but anybody who is fueled by something different, something the 1/x or the majority cannot offer them) would challenge the 1 in 1/x and usurp the thrones of the majority. Instead of a democracy of sanity, there would be an oligarchy of insanity, insane royalty and aristocrats who use the knife to stab themselves and use the bed sheets to strangle themselves instead of just going peacefully to sleep.

The logic and rational for keeping the ax out of a strangers skull would be useless and not listened to. Instead of ascribing to the manifested 1/x which society clings to so dearly, it would be erased. The conception of stopping at red lights, wiping your mouth and going to college would be obsolete.

The insane royalty would drive into guardrails and rip out their own throats with spoons, they would constantly regress to that elusive point, somewhere beyond laws and customs, somewhere beyond coffee mugs, somewhere beyond the heart of God.

XVII
Why does the world need sanity?

Humanity needs a manifested sanity because without it, there could be no progression, only an animal existence, which is what preceded sanity. However, sanity is not absolute, it is a natural occurring entity, but rather a man made aid which sometimes may need to be circumvented in order to pursue order. And order may lie outside a particularly manifested sanity, and only ones strong enough to break sanity will be able to achieve that order.

XVII
What lies beyond the x?

And so, a human participant, unwillingly drinks the fuels humanity forces onto him, he begrudgingly accepts the manifested 1/x. Yet he stands on a sidewalk, looks at reality though a telescope. An even more pressing question then enters his mind. Being forced to accept the 1 in the 1/x is one thing, but now he realizes a terrible and irreversible truth. He is forced to accept the x as well. For what is the x but the situations that humanity rejected. Some label "insane" adhere to them, but they are still human creations, defined in human words, and human conceptions. He stares at reality though his telescope, teetering on the x that centuries of the mob have carved into his skin with forks and coins. He hates even their rejections. They are all part of the same puddle.

He wants to leave humanity in piles of coffee mugs, glass and sewage, rip the buttons from his collar, crash through a telephone pole and keep driving straight into the x, shatter it and be left with whatever is there. Something he could never describe. He realizes every single thing in his life is man made and subject to the 1/x, whether accepted or rejected. His clothes, his actions (dressing shaving, driving) even his dreams and thoughts (for man has named everything he knows) are processed like cheese, wrapped and sold. Even the words he uses to describe them beyond fall miserable short, like him on the verge of transcending every thing he has ever known. Instead he crawls back to x, and is forced to stand in line with the 1. Even humanities rejections offer him no respite. But perhaps, it is here where the infinite lies. It is only when man reaches this terrible impasse is when he is able to truly conceive of the infinite.

Appendix B

The Problem of the Infinite

Can the existence of a divine entity be proved with finite ideas?

Humanity is prisoners in their own creations. They are confined in the *known*. As the late singer/songwriter Curt Cobain stated: we are "stranded by the name." And we are all stranded by the names of things, by the very syntax we use every day, by the spoken and written language that has supposedly freed us from the bonds of the barbaric. However, now, a human mind can never know anything that is unknown or that it can't name because once it knows it, is not unknown anymore. It is the paradox of civilization. Everything we create is a known, finite thing. And yet, amidst our world, amidst our problem of being "stranded by the name", there is one idea that humans have created that is even beyond our conception. Like a Frankenstein that went berserk, we have created something that we have no control over: infinity. A simple division problem of a school child can leave us in a torrent of the unknown. The square root of five leaves us with an irrational number. The elusive and never ending number of pi has baffled mathematicians for centuries. The infinite threatens us in even the remotest places.

Numbers are human. One, two and three and all our numbers are representations of the universe. They are simply names given to quantities. One apple, two chairs or ten toes, we have surrounded ourselves with our representations. To quote Pythagoras "the universe can be represented by number." Indeed he is right. The universe-our universe-is number. As stated before our numbers surround us. Four tires, two arms, a twelve-gallon tank, a thousand blood cells and trillion atoms, Pythagoras was right. The universe is number, numbers that we created, numbers that represent our world and all the finite objects in it. However these numbers go beyond the finite world because they never end. And the true paradox is, how can finite beings create an infinite concept?

The answer to this is imagination. The human intellect is in possession of something more powerful than the atomic bomb, like a million Nagasaki's. However, most are not ready for this, The imagination has transcended the mind's power to contain. However, like stubborn children, when some minds try to taste the forbidden fruit of the infinite before they are supposed to, the infinite may manifest itself in irrational mental conditions such as schizophrenia or obsessive-compulsive disorder, to quote Huxley. It is like putting an unsolvable equation into a calculator. All that shows is error. And that may be what happens when a human mind-one that is not mature-tries to conceive of the infinite. The feeble calculator-like brain fails and we are left with mental disorders and medication to try and make us "normal", to try distract us from the irrationality of our "error-ed" lives. However, if a diseased mind were to struggle though the disorders, if a mind tried to discover the last number in the ocean of infinity, it maybe left only with the things it knows. All of the irrational occurrences in the world, the evil and the goodness, the question of death, the existence of god, the depths of the human psyche, or whatever demons the mind can create but cannot accurately understand may be contained within the infinite.

Is it possible to discern a last number out of the infinity, which the human mind has created?

In this statement lies another paradox because infinity could not exist if there were a last number. However, the "last number" may not be a numerical value, as one would presume. Instead, one may have to cease using one's intellect and use ones imagination. Instead of looking for a traditional mathematical result of a strictly numerical value the answer would explode through our feeble representations and lie in a realm beyond our understanding. Perhaps the last number lies in the heart of a Christian God or in Nirvana or in some type of divine realm. Indeed, St. Augustine echoes this when he states: "All things are finite in God." Or, perhaps the last number is a glimpse into another worldly place on this earth, amidst one's daily life. However, I must articulate a fear of the great philosopher Karl Marx. He admonished Plato as well as Christian personalist thinkers for looking outside of humanity to find divinity, for holding that everything divine was not human. I believe his fears were correct. Humanity is a wicked bunch, mainly bent on self interest, all inevitably heading for decay, but the creation of infinity points to a higher realm of existence-not a separate foreign entity made to cleanse us, but perhaps something that humans have possessed, something that they have fostered since the inception of conscious thought. I will discuss this idea more fully later.

If humanities actions are viewed as mathematical equations then the last number may be the quotient to the unsolvable problems of humanities wickedness (i.e. the holocaust) or the complex and uncharted depths of the human psyche with its irrational and conflicting desires. The last number maybe the unending ever-expanding universe or it may be hell.

Pythagoras and his follows believed that the universe could be represented as number. If the universe is all number than in a sense, all of humanity are mathematicians. Every single action preformed by mankind is a sort of division problem, dividing the numbers of the universe and creating a lifetime with them. A human mathematician divides the template of the future into useable parts or quotients. However, the human intellect may not be solely responsible for the division of existence. Vast chunks of the "equation" are presumably left to higher powers to "calculate." These quotients are then further divided throughout one's existence until one reaches their death. And even death can be represented by a number-an irrational, unending number-such as the square root of 3. Perhaps the quotient to the unsolvable and irrational equation of death ultimately leads to the last number, whatever that may be.

Can a system of chaos be discerned from the abyss of the infinite?

There are patterns in the madness. English philosopher Edmund Burke once stated that sometimes all sorts of opposites mix in chaos. When this statement is applied to infinity, it can illuminate blatant contradictions. The biblical passage from revelation "I am the Alpha and the Omega, the first and last, the beginning and the end" is a prime example. In our finite world, this statement is illogical. It simply cannot be true. However, paradox is the atom of the infinite, contradiction, the building block of madness. Regarding infinity, the first and last are one. The infinite is the logical outcome of imagination. Veins of stock market trends and cigarette smoke patterns sprang from the simple act of creating numbers and counting them. When separated and "tamed", these veins become logical finite maxims. Yet, all of the "logical" maxims ultimately lead to an infinite source. Seemingly chaotic movements all adhere to this infinite.

Infinity is ultimate freedom. However, freedom without limits can lead to disaster. So the human mind has tamed infinity into oil distribution prices and radio waves.

Is infinity present in the finite world?

Infinity surrounds-indeed it threatens humanity from almost everywhere. Arithmetic computations, numbered items, even the lines in the corners of walls extend endlessly. Infinity is the "logical" outcome of imagination and intellect. All of the instances of infinity in the world form "pockets" of infinity. They are like Socrates "portal to eternity" for if one could use their imagination and not a calculator, they would see infinity flow through these pockets like a vein to the last number.

As stated before, if one were to try and achieve the computation of the last number, one would have to use one's imagination. The last number is not a quantity, but an idea. Like Huxley's "spiritual calculus", we are all mathematicians but of an abstract sort. Our last numbers could represent anything, a nice home, children, fast cars or a holocaust. From birth, one is a butcher-mathematician, hacking and counting, until death. Hacking away at existence is merely the completion of a division problem, one that has its source in the "infinite." One's last number is fulfilled upon one's last breath. Yet, throughout a lifetime, one is "up to his elbows" in hacked out hearts and things one can count, and the ability to count becomes the ability to *create*. To create is divine, but not simply to create more human finite things. The creation of the infinite becomes the cornerstone of existence (perhaps the one the builders have rejected?), the creation of the unknown becomes our purpose.

Is there a danger in trying to contemplate infinity?

A very serious danger lies in the contemplation of the infinite. In the 19th century a brilliant mathematician names Georg Cantor made many advances in the study of the infinite. He was the first person to deal with actual infinity. He developed transfinite numbers and discovered different levels of infinity. However, coupled with other mental problems and the contemplation of the infinite, Georg Cantor went mad. Anyone who tries to conceive of a never ending string of numbers certainly will feel the "delightful horror" of infinity as Edmund Burke described it. Galileo had also made some interesting discoveries dealing with infinity and was going to publish a book on the subject. However, he could not finish because the topic frightened him. He could not conceptualize it with his human intellect. So a word of caution must be given to anyone who wanders into contemplation of the infinite. It is a concept unlike any other in the world. Simply because everything in the world ends, where as infinity does not.

If there is a last number, who counts it?

Questions of this nature are beyond the scope of human intelligence. However, this may be all the more reason to contemplate them. Humanity's simple numbers as representations of quantities (i.e. six computers) may be the clue to answering this question. The human intellect has used representations for quantities, perhaps our last numbers, our very existences which we butcher and hack, which we cry over and laugh, are higher representations-like a higher order of numbers. Our "last number lives" are representations of *something*, of some *x* variable unknown to us. To echo the Jewish philosopher Abraham Heschel: "Everything known is a symbol for everything unknown." Perhaps, our lives are countable by a perfect state of order, but unknown to us.

What is the last numbers' relationship to the "infinite" and the creation of the universe?

When dealing with the last number, one must suspend most rational judgments. Rationality is merely a point of view, and one not feasible when dealing with infinites, as proved by Georg Cantor. Perhaps this "infinite" is like an ocean of quantity and space, everything and nothing. It ebbs to and fro. Quantities intersect space and wash over it in an indiscernible pattern. Yet, as seen earlier, sometimes minds try to discern this pattern; they try to "swim" it. However, the patterns can only accessed by the mind when they are simple and easily represented.

The last number is bound up in space and time; indeed, the very space human's use to perceive their world may be just another dimension of the last number, emanating from the imagination. Perhaps one has to try and conceive of the last number not *in* spatial terms, but *with* spatial terms.

How is language a major obstacle to the human mind achieving this divine knowledge of the infinite?

In the thirteenth century, Thomas of Ockham described the problem of understanding with the term *nominalism* or using names to represent things in reality. Ockham states that uttered syntax cannot accurately describe the true nature of the world. The human mind in its quest for logic gives meanings to things in the world around it. When in fact, this quest leaves us "stranded by the name." Language is not our liberator but our barrier. It prevents humans from transcending the known. Perhaps if we did not have words we could think in something else, something not as confining, possibly Plato's essences, possibly something pure and unknown.

What is the significance of the human creation of infinity?

What we have inherited from history and our ancestors is simply a battle. The second law of thermodynamics states that all matter inevitably moves towards a state of chaos, from order to disorder. And in light of this inevitable decay, what has been the aim of every civilization? It has been to build progressive, lasting structures, to surpass their ancestors, to create a better world. While there may be some disagreement as to the motivation of civilizations, what can be agreed upon is the result, which has been an irreversible acquisition of knowledge. This has led to various advancements, and ultimately, increasingly progressive societies. The very idea that humans can conceive of the notion of "progress" is evidence that we have progressed from our animal origins.

Yet, all living organisms perish, all things inevitably decay, all move toward disorder. If left unchecked or un-policed, a majority of the members of society would degenerate into madness, into roving bands of thugs competing over their own self interest. Our entire thought process has been shown to be built with an irrational subconscious. Indeed, it would seem as is all of nature were governed by some universal id, always succumbing to disorder and the irrational. Yet, we go to school, we go to work, we build structures, we teach our children. In essence, our rational progressive evolution has to submit to the second law of thermodynamics.

How does infinity then intersect with this? Infinity is a human creation, whether it is endless numbers or an endless divine being, the human mind created the notion of infinity. Like a Frankenstein, we cannot control what we have created, but, I believe its proof is existence of a bridge, a link between the two dynamics, between the two combatants of our existence, between progress and chaos. Infinity is a human disorder, a chaos where all things collide and fuse, a severing of reality where only the imagination reigns supreme. The natural state of things is decay, disorder. Once broken, an egg will never fuse together again. Progress pulls one way, chaos the other. However, the concept of infinity is a sort of destructive progression, a mean between progress and chaos. It is a man made chaos that does not inevitably decay, but rather, endure forever. And in this madness, man no longer counts, but can create his entire being.

I believe the creation of infinity is the key to our existence. Not to transcend this world, or lament the fact that all exists to die, but to bridge the gap between the opposing forces of existence, and create anew whatever we desire. And perhaps, create something higher than desire, to create something truly unknown.

What is the relationship of the infinite to members of society?

Most members of society do not care, or acknowledge the existence of the infinite. Their lives are built on a "fairy's wing." They wake up and do not see space and time spread like jelly before them, able to be recast as something different; they do not become nauseous from existence. Instead they assume that existence is concrete and stable, they assume they were meant to fill some role in the world and take their place accordingly. Instead they care only for their own skin; they care only to increase their happiness, whether that is by prestige, wealth or idleness.

Some contentious members do recognize the battle of existence; however, they assume that it is fought between good and evil, which are easily definable, polarized terms. They do not see existence as fought between chaos and progress. There is no delicate tension of opposites that rely on each other to survive; it is a knock down, broken jawed fight to the death. But these people retreat into a world of "good." They hide from reality and want to believe that if they just wait, all will be made better. They voluntarily remove all power from their hands. But "good" is not necessarily progressive and "evil" is not necessarily chaotic. Progress and chaos are terms that cease to be epistemological, but rather simply ontological. They just "are" they are not "something."

How has the "the battle for existence" raged throughout history?

Various methods of cementing existence to the earth had been tried. Massive beurcratic empires, adherence to religious doctrines, belief in reason, science and nationalism. All of these methods have been tried and have been exposed. The only successful method has been industry, business and commerce. All is now made to be assembled and sold. All except infinity. Infinity is the bastard child; the one society likes to forget. But it is the key.

Appendix C

Is the Science of Logic Flawed?

A true thesis aims to prove something. However, this thesis simply aims to enlighten. Its purpose is to force a question, a question which humanities limited intellect cannot answer at this junction. Yet, that it is no reason to abstain from asking the question.

Is the scientific system of logic infallible? Are there flaws in this seemingly impenetrable system? First, "logic" must be defined. Aristotle is commonly thought of as the innovator of logic. It was Aristotle who put "logic" on the map. While it would take centuries to take hold, its roots lay in Aristotle. He defines logic as obtaining information by using evidence. The relationship between cause and effect is crucial to the study of logic. Every thing happens for a reason, not in a Christian or mystical sense, but in a rational sense. If it rains, it is caused by specific weather patterns. The sun seems to "rise" because the earth turns. Aristotle labels the "middle term" imperative in the study of logic. One can progress to a "logical" conclusion, (such as rain) through illogical thought processes (such as God is angry) and still arrive at a logical conclusion (it rained). However, Aristotle stated the middle term, or in this case, weather patterns, defines "logic."

In this vein of thought, if logical thinking is then applied to the science of logic itself, it proves to be illogical. Cause and effect are crucial to logic. There is no effect without a cause. However, if every cause leads to an effect, then from where did this chain reaction begin? If it "began" at some arbitrary point, if one can isolate a specific incidence where there are no preceding causes, then that would leave one with a causeless effect, or an effect less cause, both of which are illogical by logical standards.

If there is no point in which one can isolate a cause or effect, then the chain of cause and effect much be on-going, eternal, and infinite for it to satisfy the requirements set forth by cause and effect. Yet, infinity is illogical; it is a term with no logical explanation. Modern physics glosses over this apparent contradiction by making this a "law." The first law of Thermodynamics states that no matter can be created or destroyed, just changed. But this is infinite at root. For if there is no creation or destruction, then it must be eternal, hence, infinite, hence illogical.

Nonetheless, a system of logic, a system of cause and effect was established. Yet, the question this "thesis" asks is whether this foundation for logic was rotten? Is the starting point for logic, an irrational idea, an exception that proves a rule? One can not say for certain. All one can do is use logic and realize it is a necessary institution for the stable functioning of society. Yet, just because a system is necessary, does mean it is not flawed.

If the Science of Logic is flawed, is it Necessary?

If the science of Logic is flawed because of a "rotten" foundation, then must it be adhered to? Has our entire existence been based around a false doctrine?

Well, perhaps it truly is the exception that proves the rule. However, even with its flawed status, logic is still a vital institution is society. As Immanuel Kant theorized, human creations, such as logic and justice, are necessary in our visible reality, because without them, existence would have no intelligible meaning. Mankind would be a pointless endeavor, swimming aimlessly in an abyss. Even with a flawed foundation, the science of logic is absolute necessary to society, it has become humanities blood.

Can the flawed science of Logic be replaced with a substitute?

If logic comes from an illogical-or at least unknowable source (i.e. a deity or the big bang), and its starting point can be proven false or at least not absolute, then can a different set of logical rules be manifested?

The paradox of this question is that if a new set of rules replaces the old logical ones, be they illogical by the present standards, they then *become* the new logical rules in the new system. However, what can be deduced (albeit with confusion and difficulty) is that the science of logic (using evidence based on cause and effect relationships and then inferring correctly) is relative and may be arbitrary and not absolute to any given universe.

An analogy can be drawn with the measure of gravity. Gravity is a relative force, which depends on what planet it is measured from, not an absolute force. Logic is also analogous to Einstein's theory of relativity. Logic is relative to place, time and existence.

Another example of flawed Logic.

Let A represent: God is the most powerful being

Let B represent: God can create anything he wants

By this logic we can infer that if God is the most powerful being (A) and he can create anything he wants (B) are true statements, then the statement "he can create a being more powerful than himself" must also be true (C). But if C is true, then that makes A false for how can the most powerful being create something more powerful than himself, he would not be the most powerful anymore. It is a paradox because either he can create anything in which case he could create a more powerful being, but if he does this than he is not the most powerful being. (the word "he" is used when referring to God only as a convenient term or amorphism to define an all powerful deity, which would most likely not have a gender).

If A and B are true, **then** C must be true, but if C is true that negates A which makes the whole statement false.

Appendix D

Obsessive-Compulsive Disorder in A.D. 2007

Despite all of the modern headway made with the disease known as obsessive-compulsive disorder, many questions still remain. Its true cause are still unknown. Not only that, it becomes increasingly hard to nail down a concrete explanation of a socio-biological disease when cultural beliefs, norms and acceptances are constantly changing. The environment in which people succumb to OCD has changed drastically from many of the researchers of the past. 2007 is vastly different time from any other time period in the history of the world, and this undoubtedly will affect a mind vulnerable to OCD. A question which begs an answer than is left on our lips. What is the state of OCD in 2007? In this rational, ordered, devoid of spirit, consumption driven, individualistic society, how does an absurd disease fit in? In order to answer this question, one must inevitably return to the past to discover the attitudes and perceptions that prevailed and then contrast them with the new factors influencing the diseases today. However, one must be very careful to avoid mass generalizations of society and realize the diversity of circumstances in one era of history.

During the middle ages, there was a strong, almost childlike dependency of society on the Catholic Church to lead them to a better life in the form of heaven. This "backbone" of society began to give way in the mid 1300s due to corruption in the church, failure of the crusades, the brutality of the inquisition, the Black Death and false teachings. During the Renaissance, man still believed in God, but realized his religion was flawed and began to branch out. Following was the age of Enlightenment, when science replaced religion, and reason replaced devotion. Man thought that science and reason would carry him over the threshold, where he once relied on religion. Yet, with the advent of quantum physics and it's unknowable and unintelligible outcomes, the study of the infinite, the belief in evolution degrading humans to almost animal like status, the horror of the two world wars, and lastly the terror of the atomic bomb, science and reason were exposed and shown to be as faulty as religion. While these are general and vague assessments of the last three millennia, and may not apply to all segments of society, and in fact took many centuries to come to fruit, most historians could probably agree that it is a semi-accurate account.

This leaves us in 2006. While in the last three millennia man had something to believe in (and many do still believe in those things) they have been exposed of their faults. In an almost nihilistic sense, one could say that the institutions of religion, science, reason and nationalism failed the west. So what does modern man have to believe in? While some may not agree with this assessment, still others may not care, others still cling to the past, and one still has to ask themselves where society is at. Unfortunately there is no answer, in fact there may not even by a question. But for those ambitious ones who try to asses the situation, either consciously or unconsciously, some may find themselves succumbing to mental illness. As Foucault states, now, maybe madness draws the contour of the void where all of these other things failed.

In the last three millennia, the western world has undergone many transformations. While these transformations were experienced differently for individuals, most historians could agree on a "collective feeling" of the age. And these collective feelings undoubtedly play a role in mental diseases, especially OCD. Since OCD is a biological as well as social disease, in each era it has manifested in, it has also comes with different social meanings. In the middle ages and then the Renaissance and Enlightenment, it may have manifested along religious or scientific lines (and obviously along many other variations, depending on the individual.) However, 2006 is a much different social atmosphere. As Adlous Huxley once wrote of mental illness, they [mentally ill patients] may be trying to conceive of a higher reality but cannot express it in their limited framework because while there was religion or science, now there is nothing. So it has become almost near impossible.

The modern age has also been fittingly referred to as the "age of anxiety." And so, like the appearance of "error" on a calculator, or a frozen computer screen, some feeble minds already prone to disease malfunction, similar to Sigmund Freud's analysis of an obsessive mind (which stated that an obsessive mind "gets stuck" between conflicting desires and norms), and than the sufferer can only obsess over the things *it does know*, be it eternal damnation, fears of death, the raping of ones mother or germ contaminations. So, Huxley's observations that "perhaps mental illness was simply the individual's failed and fatal attempt to conceive of the divine" may still prove useful in the modern world. Huxley was mainly referring to schizophrenia, but the general idea is the same. Mental illness is something in which an individual loses touch with reality. Whether the person believes their own delusions, or knows their absurdity but continues to have them, they deviate from the socially constructed norm in which they were born into. And for three millennia the reality in which the vulnerable mind had to return to was a solid one, based on theology or science. However, that has faltered in the twentieth century. So, perhaps in a perverted way, these obsessions and compulsions now are the only things able to fill the vacancy left by the failures of the past three millennia; they become the only thing for mind to truly believe in.

Such vast sweeping and mystical diagnosis are usually frowned upon by traditional medicine today. However, traditional medicine and the understanding of mental illness in the sense we know it in, has only been truly fruitful in the last hundred and fifty years or so while people have suffered from mental disease for centuries and needed explanations. And while Huxley's poetic view may not be "rational," and may not easily fit into the structured framework of modern science and medicine, or may be painfully reminiscent of the older modes of thought, the mental diseases in which we face today -especially ones like OCD- are anything but rational and structured, and refused to be truly understood. In his great work *Madness and Civilization*, Michel Foucault wrote: "In the modern world…modern man no longer communicates with the madman." The lunatic is usually seen as an abnormality, a stain on the rational fabric onto which modern society is supposedly built. However, that language, the language of the madmen, the absurd obsessions and irrational compulsions, after three millennia of failures, may be the only truth we can know is real.

Appendix E

Hell on the Kitchen Table

I

Stains on the paper made it barely legible. He scribbled his blasphemous thoughts through the blood. He never stopped; he rarely ate or slept. Alone in his shack this modern philosopher, this reject of the contemporary world drafted his treatise of sorrow to the audience of himself. He was alone, yes, but with an undeniable purpose.

The books are stacked so high in his apartment and they seem to watch him. They watch him scribble, they watch the crumpled papers pile up because nothing he writes ever satisfies him. The sweat collects on his cramped fingers and his heartbeat accelerates like a machine gun. Now is one of those rare moments. The thoughts and ideas perform. The lonely audience of his fingers applauds.

A picture of Jesus hangs above his table but he can't stand to look at it. Actually he thinks the picture can't stand to look at him. And the thoughts never stop. That's why the picture hates him. The evil thoughts spin like a centrifuge in his skull. Every horrible vision ever conceived is his. Like some demon mother birthing a thousand insanities every second for him to enjoy. And that's why he has to write. He can't look at that picture. He just writes until his hands ache and curl.

II

"If death is an absence of life, then life is merely an absence of death. A state of functioning in which death is simply not active."

Oh god the thoughts raged. Blasphemous things! This is not Christian doctrine, he was not taught this as a child! These were his thoughts. He negative terrible ideas.

"In the absence of death a type of conscious spawns, it is wrapped in a physical body which is nothing more than a weak tangible structure laden with desires, with faults and unhealthy processes."

These were not God's words. These were his own twisted words.

"The consciousness appeases the physical by acquiring the desires that is precipitates. It is forced to or the uneasy harmony of the thought and the feeling will be disrupted. The singular body wants nothing more than but these unhealthy desires, it cares nothing for a range of reason or speculator genius. And yet I ask, is the body better off than the consciousness? In this absence of death, and death which is permanent and inevitable state of things-would it not be better for the body which has the advantages of tangibility and movement and placement and being, would it not be wiser to reject the lonely conscious and adhere to its cravings?

Probably not but it should be considered that genius and reason have isolated men amongst their peers, genius and reason have persecuted and tortured men because of an adherence to virtue. Conflicting ideologies have torn nations apart on their human seems. They have left lands bloody with dead men and lonely populations. Pious brainwashes have slaughtered weaker nations in the name of knowledge and reason. And so I ask this, instead of our preoccupation with limited conscious, would it not be wiser to concentrate on our own human bonds and on the desires that precipate them?"

What did this mean? The picture of Jesus stared ruthlessly down at him and his own guilt. The philosopher or lover of wisdom was condemning the very knowledge he used to write his words, the very values that shaped life. But what life is it that is shaped by a blind and answerless faith?

III

He was taught to believe in God out of the strictest fear. A fear of damnation and hell fire. He was taught to adhere to every rule and doctrine. Taught to reject all the desires and questions that his creator had given him. But then he started to write.

"And what kind of cruel joke is it, what kind of devilish ploy is it to create life from nothing, to rise something out of its natural absence of life, to create two conflicting entities- a consciousness and a body under one skin, controlled by one brain. To fill the conscious with rules and fill the body for the utmost capacity for breaking them"

Now he felt like a blasphemer. He was a disgrace to his family that held him in such high moral esteem. He was supposed to be their good son, their moral champion, and their finished product of perfection. He was supposed to have faith, believe and never question. Never, ever question. But the questions released themselves from this demon in his skull and he was powerless to stop it.

"Why create a conflicted thing with the easiest capacity to sin, a thing which had no voice in its own creation and then damn it to eternal suffering? If I had had a voice in my own creation or any power to determine whether I existed or not, full well knowing what kind of struggle I would be locked in, I would vote no, I would choose to remain adrift in the void of non-existence. I would simply want to be nothing."

IV

"I am not a condemner of wisdom, surely we would not be better off as slobbering savages, eating, mating and dying, but we would undoubtedly be better off without our good intended knowledge, twisted into the most sickening perversions that we label wisdom."

And this is when he felt the worst. This is why he pounded his hands until they bled, to alleviate the intense pressure from the guilt. Everything he wrote he felt went against what he was taught, what he was raised with.

"Twisted perversions such as love. Knowledgeable men who sought to label and define the bonds between two human beings, who were not wise enough to realize the only true bond between humans is one of flesh and pleasure. Men of wisdom who did not want to admit that we as humans are maybe not capable of "loving" each other. Maybe we as humans are at least smart enough to create and define a bond between us. However we are unable to follow it through. True love is an institution that drives the world and its populations to pursue a falsely attainable notion. It is a notion that a human animal cannot possible achieve and yet has the false belief it can.

Love cannot be defined as anything more than a glorified lust that can be started, annulled and re-started again. Love to humans comes in spurts, it blooms in short seasons on altars, in bedrooms, it dies in courtrooms, is reborn in another bedroom. It is cyclic, revolving us non-dead beings into different states of sensation. We have the capacity to define it, but not the ability to achieve it."

He was disgusted with himself. Basically he believed that love was nothing more than chemicals in the brain and skin. He reduced God's greatest creation to animal desires and secretions.

This apartment was his cell. A timely cliché. But a clichés ceases to be a clichés when they are true. The rickety table where he sat was a podium. He was the lector at his own funeral. He wrote the eulogy that no one will care enough to read. The words scribbled on the papers in front of him were his requiems. He condemned himself further with each blasphemous word. There was no relapse, no pardon from this sentence. The warden hung in a picture frame and had a burning heart. Jesus stared from behind the glass. He felt guilty, his verdict was guilty. His crime was of skepticism and blasphemy. Everyday he single-handily destroyed more of his faith. He never questioned but then he began to read and to think. He searched the work and his own mind. He crawled from the confines of his own skull into a slightly larger cell of the earth.

Appendix F

I
Roughly 6,000 BC, Catal Huyuk, (Modern day Turkey)

A small village of roughly 500 inhabitants thrives in what is today modern day turkey. Crops are grown, goods are exchanged, and technology improves the standard of living. Elders share their history and stories with the younger members of the tribe. In this village, the seeds of humanities bustling society have begun to sprout, in all of their forms.

"In the old days, when I was a boy, before I came to live in this town, in a time when my family still chased the wild beasts for food, my tribe had finally made peace with a tribe from the mountains that we had fought with for ages. For a while, there was peace, and all was quiet. Then, one night, the men from my tribe snuck into the mountains, and since they had visited the village in peace time, knew exactly where to go. They found the men sleeping in the huts with their wives, the same men that had sworn to peace. The men from my tribe bashed all of the men's skulls in. They raped their wives and tore some of the children apart by their arms and legs. They then stole all the valuables and food and came back to our village with it."

The listeners were silent. Some had eyes wide open, and waited for him to finish.

"These men had shaken hands; they had traded with each other. They trusted one another." He said quietly. "But later, the same men bashed their skulls in."

The usual silence followed Shibo's story. The listeners, both young and old, tried to conjure a picture of the grisly scene. The only movement was a Tiko, a young boy, who tended the fire. The burnt wood smell emanated from the fire pit.

"Why do you sit and tell stories of the past old man?" Mitubo smiled at his grandfather, breaking the silence.

His grandfather, Shibo, smiled. He stoked his grey beard. "The past," he began "is what we are."

"Well, maybe" Mitubo said "but-"he pulled a handful of sharpened flint blades from his pouch. "I don't like the past so much, I like now better."

The blades were oblong and jagged. But they were sharp. Sharper than any tools that Shibo, or anyone else around that fire had ever seen. Shibo's story was quickly forgotten. The people around the fire crowded around Mitubo to see this sharpened blades. The young children grabbed greedily at the blades, but Mitubo kept them out of reach.

"How did you make them so sharp?" a younger boy asked Mitubo.

"It's my secret" smiled Mutibo.

"You should make really big ones." Another boy said.

Mitubo looked in the direction of his grandfather; he then looked down at the boy and smiled.

"I am working on it." he said

"You always seem to be working on something," another man Mitubos's age said to him.

"Yea Mitubo, you always seem have some new thing." Tiko said. "I like the wet colors you gave us. I color on rocks with them. You are so smart."

"It just ground up plants mixed with water and animal fat" Mitubo smiled, but he said it arrogantly, as to send the symbol to the other men, as to take credit for thinking of these new ideas. Mitubo passed the blades around and watched smugly as everyone heaped praise on it.

Shibo, still smiling, slowly rose from the rock he had been sitting on. "You all forget so quickly. Well, it is time for this old man to go to sleep." He patted his grandson on the shoulder and walked over to his hut, unnoticed in the commotion over the blades.

"Goodnight old man." Mitubo whispered

II
The Next Day

Mitubo carefully arranged the items in front of him; the new sharpened flint blades, colorful dyes, beads, hooks and containers and animal skin jugs. He looked on with a sense of pride. He had created all those things. Maybe the gods had inspired him, but even if they did, he was the one who actually thought of them. He was the one who would take the credit. For many of the people in the town, life had been made easier because of his inventions. He dipped his fingertip into the dye and smeared a portion on his rock wall. Kavala, a stocky, well built man, appeared at the entrance to Mitubo's dwelling.

"How are you?" He asked Mitubo "Any new items?" he asked in jest.

"Not yet," Mitubo held the sharpened flint blades. "I'm still working on these"

"Still working?" asked Kavala. "How much sharper to want to get them?"

Mitubo gave him a dubious smile. "Not shaper, bigger." He said.

"How big?"

"I want to put huge sharpened flint stones on the edge of our spears. I want to make axes with them also"

Kavala took the small stones and gently pricked his finger. He watched as the tiny point drew the skin of his fingertip upwards, when he released, a red blotch instantly formed.

"This will make us unstoppable." Kavala returned Mitubo's dubious smile.

"Now we just have to find someone to use it against."

Kavala laughed. "There will always be somebody to fight."

"And if there isn't," Mitubo said "we'll find someone."

* * *

Shibo sat quietly in his dwelling. The stone walls made it surprisingly cool inside the small space. He had been up before the sun rose. Thinking. Everything he knew would be gone soon. Soon, he would be like the stories he told, just stored up in some old man's brain, like a rock container. Everything ached. His legs were weak, his arms thin, His skin pale. He would not live much longer. Shibo believed that when he died, he would be taken back into the loving arms of the mother goddess. At least he hoped he would, because if there was no mother goddesses...well, there had to be, who else could have created the earth and made the ground so fertile?

Shibo painfully arose, dismissing his doubts. Questioning of the mother goddess was not for him. In his youth he was a hunter, then later a farmer. Sometimes he missed the old days, he missed the thrill of killing, but he had settled down, once the elders had learned the secret of the earth, had learned how plant and harvest, he adopted their ways. Most of his life he had lived in caves, but it was nice to settle down in one place. Small circular dwellings were constructed, strong manmade things that could resist the heat and cold. Now he would die peacefully in the same place he had lived for many sunrises. He watched as his small village grew, all in his lifetime. And now, that his body was failing him, he served out his sunrises by using his ideas. By telling the great stories of his youth to those who would listen. That was his purpose now.

He walked out of his dwelling to the granary. He liked the bread in the morning because it was warm.

III

"It is a call for war!" Ketnaga stamped his foot. A small dirt cloud arose. "They want our grain and water! They won't stop until they have destroyed us!"

Mitubo looked gravely at the warriors who were clambering for war, and then at the elders. "I agree." He was the unofficial head of the warriors, and spoke for them.

"I do agree that the Zemsi might harbor hostile feelings, but we cannot simply attack another tribe." Mevessa, one of the most respected elders replied to the agitated warriors.

"If we wait, we will die." Kavala said bluntly.

"First, we should pray to the mother goddess, and see what she has us do. Two priests should burn handfuls of grain and slaughter a goat, interpret the symbols, and then proceed from there." Hala, an elder, suggested.

The many warriors that had shown up at the council were outraged at this. Their penchant for war unnerved Shibo who watched silently as the council and the warriors deliberated. They stamped around no better than the goats.

A wandering tribe, called the Zemsi by Shibo's people (who called themselves the Kiatga), had been a constant threat for many ages. One Zemsi hunter had attacked three Kiatga hunters a few sunrises ago far way in the plains. The hunters came back and said this was an outright call for war. The elders however, thought that it might simply have been act of bad miscommunication or perhaps self defense. To rush into a battle was not a wise thing to do, lives are lost and things are destroyed in battles, but the warriors of the Kiataga tribe were eager to prove their strength. Shibo stroked his beard and waited to speak until the voices and lowered.

"Warriors, elders," he spoke in a calm manner "we cannot rush into a battle. We have too much at stake. We have a village, we have young children and pregnant women. We have fantastic structures and food supplies. The Zemsi are not an evil tribe. I suggest, we send an offering to them. They are skilled hunters, better than we" at this some of the warriors became agitated "perhaps we can learn from them. And we can teach them the ways of the earth. We can trade goods with them. We can learn *from each other."*

At this, the warriors were outraged. They raised their clubs and demanded a fight.

"How can you say that!" Mitubo raised his voice. "They eat their own children! What makes you think they won't do the same to us? The Gabon hate them as well, they have agreed to help us fight them."

Some of the women became frightened at this. Shibo, sighed at the ignorance of his grandson.

"I see you warriors with clubs hoisted; ready to bash the skull of anyone, no matter whom. You speak of the Zemsi as if they were no better than goats. Have you ever spoken to a Zemsi? I have. I know their people, as many of the elders do. They are fierce in battle, but they are not animals. And you speak of the Gabon tribe. The Gabon are the ones to be feared. And yet you so readily want to join them. They are the ones who rip the babies out of pregnant women because they think it delights their gods. Men are men, no better, no worse. I urge the council to send and offering, not make war."

The head elder of the council, Marxum rose to speak.

"The council has decreed that war is not a good option. We will take the advice of Shibo, and send an offering."

The warriors once again raised a cry of disgust. But Mitubo did not speak right away. He the addressed the council but his gazed remained fixed on Shibo.

"Mark my words, you will regret this. Our entire village will." And then he and the warriors marched off.

The council did not heed Mitubo's warning, but rather set plans for a peace offering to be made. However, Shibo was still unnerved by his grandson's actions and demeanor. Why was he so eager to fight?

IV

"We cannot make peace with these animals" Kavala said.

Mitubo played with small blades. "Yes, we can, and we will." He finally said disgustedly.

Kavala looked at him with surprise.

"Of course there will be peace. The Zemsi are not the demons everyone makes them out to be." Mitubo walked to other side of the partition in his dwelling. He removed some baskets to reveal a flint tipped spear. Kavala looked on in awe.

"Peace is for women." He gently stroked the flint, realizing its sharpness. "Those idiot Zemsi will agree to peace and we will not be able to use my new spears. We will not be able to take their food."

Kavala looked at Mitubo awkwardly. "The Gabon want to help to, with our strength and the Gabon, we can surely defeat the Zemsi–

"The Gabon are fools." Mitubo cut him off. Once we they help us finish the Zemsi, we will turn our spears on them.

"You just said that there was not going to be a war?" said Kavala.

Mitubo violently stabbed one of his plant reed baskets in behind the partition. Kavala started. The flint sliced easily through reed and made a soft clink when it hit the stone floor below.

"Not yet." Mitubo smiled.

V

Young and old gathered around him. Mothers, children, even some of the younger warriors. Shibo sat on a raised stone and began to address the villagers.

"We have decided to send a peace offering to the Zemsi." He said calmly. They are a peaceful people, just misunderstood.

"But one of them attacked a hunter!" a man cried out from the crowd.

Shibo looked at the man before addressing him. "Tenna," he spoke calmly "you speak as if you were there. One lost hunter attacked three of ours. That is what happened. Why would one man plan to attack three? Why would he throw one spear and then run away? Too many details do not make sense. And we cannot fight a battle on nonsense."

Tenna shrank back, he seemed to have accepted what Shibo said.

"But they have always hated us!" Rassa, a woman shouted out. Some of the crowd agreed.

"And so that gives the right to destroy them? Because they hate us? Do you think we are that much a part of their daily lives? When they are hunting deer or gathering berries, do you really think they have us, the Kitaga, in their heads? Do you think they stay up past the sun rise hating us? No, I think they go to sleep and make love to their wives without thinking about us."

This joke lightened the crowd, some even laughed.
Rassa did not answer.

"We hate them and they hate us, and the hate just grows further until one destroys the other. If one stops hating, then the other has nothing to hate, and all can live in peace."

The crowd seemed to accept this, when all of sudden Mitubo broke the silence.

"And you think by loving someone that hates us that will end the trouble."

"I wish it were that simple," said Shibo, "but I do think if we let go of this hatred and perhaps tried to understand them, things would be better."

"Animals in the jungle do not try to understand their prey. They eat them and are safe." Mitumbo said

"They are safe until a bigger animal eats them" Shibo said.

"That is why we will team with the Gabon and destroy the Zemsi."

"You place your faith in the Gabon, you place your faith one set of men against another. But what you fail to realize is men are men, no better, no worse. And I wonder if you are no better."

This drew a look of rage from Mitubo, some of the women in the crowd backed away. Mitubo's fists clenched tight and he stared down his grandfather. But his grandfather continued speaking.

"You are young. You have not seen what I have seen. I have seen men slaughter men in battle, then wrestle an animal to save a new born baby. We are all evil and all good. No one is completely good, nor completely evil. What is needed is not blood and war, just understanding."

Mitubo sliced his forearm with the sharp flint spear. No one had noticed it will the two men were yelling, but now all eyes were drawn to it, as it sliced the top layer of skin open on Mitubo's forearm.

"This" he smeared the blood on his face "is the only thing that matters. Each tribes hates each other, one will live and one will die."

"I wonder if another reason you and the warriors are so bent destroying the Zemsi are because they roam the plains where we hunt as well, and they have good soil." His grandfather said, ignoring the blood.

"I care not for soil, or others hunting grounds, I care only for the safety of my people, something you obviously do not care about."

Some of the crowd gasped at this insult, but Shibo remained calm.

"I wish things were as simple as you make them. Such as destroying them in the name of safety. You manipulate people's hate and use it for your own needs. It is easier to hate and fear because you do not have to think, just kill."

Mitubo gazed unflinchingly at his grandfather. This man had taken care of him when his father died. He owed him much, but, their personal allegiances died now.

"Your stories cannot make us safe old man. Only my spears. Go lock yourself in your hut and wait until the battle is over."

"There will not be a battle. We have sent a peace offering and the Zemsi have accepted."

"For now." Mitubo said. He jabbed the spear into the soft ground and walked away. The crowd gradually dispersed.

VI

"She is with mother goddess." Shibo overheard their conversations. Trialao combed his dead wife's hair and began to wrap her in animal skins. A section of the floor of his dwelling was removed and ready to be the final resting place of his beloved wife.

Griota, an elderly priest, approached Trialao. Some of the townsfolk gathered around for the final prayer. All bowed their heads and waited for Griota to speak.

"And now, mother goddess; please accept Tippa, beloved wife of Trialao." He gently touched Tippa's lips and Trialao pulled the animal skin over her face. Another man helped Trialao place Tippa under the floor of his dwelling. When she was placed correctly inside, Griota threw some water on the corpse and the men closed the slab and it became the floor of Trialao's dwelling again, but this time with the body of his wife underneath it, forever.

"And Tippa, be still in the afterlife, be happy there, your body will remain under the dwelling of your husband, so you can still be close to each other in death as you were in life."

Shibo watched as they buried the body of that sick woman. He had known her well, and now she was dead. Just like people would one day watch him be buried under someone's dwelling. Death crept closer every sunrise; the mother goddess hopefully prepared a place for him in the next world.

He often wondered what would be in that next world; he wondered where Tippa was right now. Tippa's body had been prepared in the body house, and then buried under her husband's dwelling. And then all would forget and go on hunting and planting. But he did not forget, he never forgot, he wanted to know. Some priests had even said what you did in this life determined your place in the next, but he was never sure. All he knew was that his life was uncertain. He did not know what would happen when he died. All he knew was what had happened before.

VII
Well After Sundown

Kavala stealthily walked past the huts. Shibo could not tell why, but in the moon light, his face looked darker, much darker than his skin. Two other men, one of the warriors and Mitumbo, also walked with him. Shibo had not slept well in years; the slightest noises woke him up. He watched through his window as the three men crept back into the village. They all carried a weapon, presumably on of Mitubo's new sharp spears. While it was hard to tell in the moonlight, the men seemed proud; they walked close and almost looked like brothers. And then Shibo realized what was on their faces and hands.

Blood.

Shibo gasped to himself. They did not see or hear him. Kavala and the other man went to their huts, but Mitubo stood a moment, staring at his grandfather's hut. He then quickly walked in that direction. Shibo decided to meet his grandson head on. If this were to be the end, Shibo would meet it bravely.

Quietly and slowly, the door opened. Mitubo stood in the doorway, covered in blood, holding a flint spear in one hand and a large animal skin pouch in the other. Without saying a word, Mitubo emptied the contents of the pouch onto the stone floor. At first Shibo only saw blood, but then body parts fell from the pouch. Fingers, noses, ears, and a babies' entire arm.

Shibo, equally as stoic as grandson did not say word. He remained strong.

"I told you I would make a war, you foolish old man. A few men and I went to the Zemsi resting grounds and killed two families in a cave. The Zemsi men will find them in the morning when they awake. We left an amulet so they will know who did it. When they will bring their army and attack us we will slaughter them."

"Your weapons do not make you right." Shibo said

"Yes they do. I am sick of listening to you and your stories old man. No one cares about what happened ages ago. I respect you for taking care of me, but there is more at stake now, there is our village."

"There is only your army and your spears" Shibo said, almost with a half grin.

"You have figured me out old man. But you will tell no one. Sometimes things need to be done to protect the people."

"You aren't protecting the people. You are only protecting your own interests. But I'm sure you already know this."

"And now, no one else will." Said Mitubo. He thrust the spear into the bottom of Shibo's jaw. The flint easily pierced through the skin, through the tongue, through the top of the jaw and into the brain. The tip emerged at the top of Shibo's skull.

"Now, you won't tell anymore stories, even in the after life, your jaw won't move, I have pinned it shut. The past is dead. Be silent old man."

Shibo quivered as he died at the end of Mitubo's spear. Now, the past was truly dead, and the future could continue on as planned. Mitubo dropped the spear and Shibo's body made a thud when it hit the stone floor.

IX

Mitubo and his warriors stood ready. The town had been mobilized and made ready for war. The sickening war cry of Zemsi echoed as the army approached. The women and the children had been warned. Everyone in the village had been told that the Zemsi were attacking, and indeed they were. There warriors stormed down, ready to kill Kiataga, ready to avenge the death of their slaughtered families, which only three members of the Kiataga knew about. The rest just knew that the Zemsi hated them, and that was all they needed to know.

"Warriors raise your weapons!" Mitubo screamed. The men raised flint spears and giant flint axes. Some of the men had blow guns loaded with sharp tips. No one noticed that Shibo was absent.
The shrill of the approaching warriors became louder and louder. Kavala's face glowed, his eyes widened and no one noticed he was already covered in blood.

* * * *

The battle was quick. Zemsi warriors were strewn across the ground. The Zemsi had no idea about the Kiatagi weapons, it was not a fight it was a massacre. The Kiatagi raised their weapons in victory. Some bodies were filleted, bits of skin fluttered like confetti. Blood was everywhere, everywhere. Zemsi and Gabon were dead. No distinction could be made. The Gabon had waited nearby, when the battle began they helped to slaughter the more numerous Zemsi, then as planned, Mitubo instructed the new spears be turned against the Gabon who had "attacked them" in the heat of the battle.

"We are victorious!" Mitubo's voice echoed in the village. He eyed a still breathing Zemsi, quickly approached him, and stabbed a spear through the top his skull. The spear easily slid to the ground.
"We no longer have to be afraid! We are safe! The Zemsi are dead! The Gabon turned on us in battle and we have eliminated them as well!" The blood began to cake on his skin and lips.

Frightened villagers looked at him as if he were God.

"My only regret is that Shibo, my grandfather could not see this day. He was mowed down in the battle." No one bothered to ask where he was earlier, or how he died, or why he had one of Mitubo's spears jammed in his jaw. "As is the custom, I will bury him under my dwelling. But rejoice! Victory is ours! Safety is ours!"

And they were safe now; their great threat had been eliminated. Mitubo's weapons had saved the day. Yula, one of the warriors looked wantonly at all the jewelry and weapons on the dead bodies of the rival tribes. He thought of the land the Zemsi roamed. He wanted to harvest their grain and trade it. He thought of all the things he could do now. Mitubo instructed the women and the teenage

Appendix G

The Poetry Man; Letter to the Unborn

I

It is ironic how I ponder St. Augustine's words about the creation of the universe the very day I found out that my wife is pregnant. A new soul, an unborn lifetime grows in her belly and at the same instant I wonder where *all* life came from. Augustine's words are ambiguous, unclear and confusing, but I can glean some ideas from them. Was our baby's soul kicking in the primordial soup of the universe? Splashing bits of matter into no where because it did not exist yet? And now the baby, the life, the soul, the spirit that is part us has embarked on a journey and agrees to be born. But we did not consult it! Am I no better than adam? then god? I did not ask if it wanted to be born. I did not ask if it wanted to live and abide by the contradictions of this world.

II

Once I had a dream that my grandfather killed a man. He shot the man in our front yard. People swarmed around, they fussed and screamed, the police were on their way, but he, my grandfather, did not care, he remained calm, and happy. He had done God's work by killing a man and would now gladly pay the earthly penalty, because he thought his reward would be in heaven.

And twenty years later I had a sickening premonition that he had killed me.

And it was permissible because maybe I was an enemy of God. But I asked him, if God knew who the enemies and the sinners were, than why create them in the first place to suffer eternal damnation? He said because nothing, *nothing* can prevent God the infinite from doing anything. If he did not create the sinners, then something would be greater than him to prevent him from doing that. And yet, I answer to this, nothing prevented the inquisitors in the middle ages from preventing the life of heretics. Nothing prevented Torquemada and Isabella from killing the Jews and Muslims. Instead, the heretics were burnt at the stake. Yet, I was told that God could not have prevented the sinners from sinning by preventing them to live. But man did.

110

By *this logic,* man is superior to God.

And this is why my grandfather will kill me.

In the beginning there was wisdom and wisdom in the beginning. Eternal wisdom much greater than our own. From the wisdom, God created the universe by speaking, and now I await your creation in nine months. I pray that you bring some of that wisdom with you.

III

The Dao is present when the horses haul manure, I am the poetry man, signing the internal verses of an indescribable notion, a nothingness that pervades and becomes all, a zero that becomes infinity in the midst of daily life, among the commotion of traffic and changing classes, a baby will be born, a lyric, a word, poetry in skin,

In the beginning, there was the word, but now there are just speed limit signs, test scores, auto body shops, Chinese food, touchdowns, earrings, linen bed sheets, the world, the universe, a spiral notebook, and yet...

A baby dwells in the mothers womb, a universe onto itself, heart and lungs nestled within a grain of rice, rice in a patty in china somewhere, maybe laozi dwelt among those grains of rice long ago, now they are symbols, history is symbols, I am a symbol...

Of defeat, of a broken helpless confused man about to be a father, learn from my mistakes, my verses have released me, I am the poetry man who won in defeat, because poetry sings above the hum of my computer, above the frequencies and wash machines, above the gas mileage, poetry is beautiful and infinite, logic cowers in front of infinity but poetry *is* infinity because in the beginning their was the *word, and the word dwelt among us and became flesh...*

But what about the Buddhists? The Taoists? My Grandfather says they are all going to hell if they do not accept Christ, but I do not believe this (and this is why he will kill me), they believe the earth is eternal, existing forever with no beginning, (much like the Christian God) and so both religions unconsciously agree on an eternal substance (universe and/or God), both have a destination, Christians to a heavenly paradise and Buddhists to nirvana, a nihilistic extinction of the universe yet also a non-extinction, something between the two states, Christians want infinity, Buddhists want zero...

Is my unborn a mix between the two? 0 + the infinite in an equation of unbounded existence? Unbounded existence that implodes within the primordial soup, that fills the zero like a water balloon, held in the hand of God? Maybe it is not existence or not non- existence but (x), the unknown Buddhist conception, a fraction of 0/infinity, my unborn is formed from the logos, from the words of God's wisdom and wrapped in skin, grown in my wife's belly, in the receptacle of reality.

We are not infinite, we are not zero, we are not heaven or nirvana but just some "rational" number, perhaps 25 years old or twelve miles to the gallon simply because I cannot answer the question "how was the earth created" or " what happens when I die?", we are the finite numbers between zero and infinity, like little Zoloft pills swallowed but unable to calm,

But some of us write poetry instead of count, and that is what I am writing for you, unborn, the poetry of the universe, gods numbers, the beautiful irrational in an incoherent thought, the absurd gift of a deranged father.

IV

What will you be born into? What type of place is this? Some have said it is scattered, the ten thousand things of the universe are all separate and dislocated, adrift in a sea of anxiety, able to be piled to the sky, billions of objects, trillions of *things,* miles of stuff. But others have said the universe is simply one, the scattered objects are simply an illusion, and they are all somehow cosmically joined despite their seeming separateness.

This used to be a world of Gods until logic destroyed them. There was a proposed compromise but logic conquered all of them, beat each god into submission, choked the gods until they crawled back into the netherworld, and governed the earth as a dictator. Logic has banished anything it cannot control; infinity and death are "abstract terms," that have no real meaning anymore to anyone important, even Jesus and Buddha are subject to logic's tyrannical reign.

But the gods are not dead! They have been tamed and banished, forced into line like Alaskan snow dogs, but they live! Aphrodite sat in the church as I married your mother, Ares lives in Iraq, George Bush is Horus, Marduk is the big bang! But it is Seth that thrives. Logic tried to kill him but he is only hibernating, I saw Seth the other day. A turkey vulture dragged the corpse of a squirrel across the street. And there was Seth, smiling, hibernating in the animal world, ready to thrash logic and drag it like hectors corpse, drag it across some cosmic street like the squirrel. Seth is disorder and chaos and violence, he thrives in men's hearts! Where else could he live? We hurt and kill and rape and murder, we destroy and have destroyed for millennia, we burned heretics in the name of god, Christians have been torn apart by dogs, Jews gassed, Seth cannot be beaten, he only grows stronger, he lives in the dark because we are afraid of the dark, the dark corners of the universe where logic refuses to go, the dark forest at night, logic is weak and its days are numbered!

What are you? What will you be? A human being, a being that somehow lives, and is animated by an unseen force, yet still an animal at its core, an animal that needs to be fed and that needs to mate and kill to survive. 8,000 years ago the animal-man learned how to feed himself, learned how to domesticate plants and procure a stable food supply, it was only when the animal was fed that it could settle down and create its masterpieces, its governments its holocausts and inquisitions.

Who has made you? Who are your parents? What is a parent? Oh, my little one, we are lonely souls, we pined in existence until we found each other, your mother is beautiful and sad, a lonely blond face and beautiful, you grow within her from my seed, you are *us,* but not our loneliness, not our pain, not the ground teeth and clotted mascara, no, you are something more than us, the whole is greater than the sum of its parts somehow, in some mystical way, you are part god, you are nourished from her blood, and her blood is power. It survived the loneliness cancer that almost ravaged her, your mother and I shared this lonely existence until we loved each other, until infinity and zero met in a cosmic embrace, a dance in the stars, but my sad thoughts are blood, and her loneliness is there, but not for you, never for you,

Advice: you must unlearn the universe; unlearn the space and time in which you live, unlearn the color and language you speak, unlearn your hurt, sorrow backwards is illogical and so is all your will-be knowledge. The unlearned is poetry, it is infinite, like the mariner on the wide, wide sea, but the infinite was there, lonely, yes, in the moonlight that I told you to forget, but how? You will spend your whole life building and learning, learning why that moon glows, learning the quantities, the lines on the paper, and learning how to pray so…

Learn to unlearn!

Learn the poetry of nothingness!

Learn to count with rejected numbers, negative height, imaginary I.Q., transinfinite waistline, build a being with carpet patterns quarks and strings, place the being in the anti-Cartesian plane, the one he abhorred (as well as the Greeks) illogistics, you must unlearn zero and vomit infinity across the dateline yesterday somewhere in the sea, 24 hours before I wrote this to you, but

This is unpractical, so learn the ways of the world, though they be temporary, we compete and insult but do not make better, like your great aunt, she is the aspiration of the world, beautiful and rich but logical, worthless, mortal and a taker. She sucks the logic of the world and licks the bone of the universe, she prances around her Benz, chasing others who are more powerful, anyone who can feed her logic fix, her craving eats her heart with expensive silverware on thanksgiving (grandma wasn't invited), your great aunt may well as be the turkey, stuffed and bent back, tied with strings, head chopped off, she wants you to look at her, she wants you to eat her like a medieval Jesus, but you can't, do not sacrifice her, light the gravy on fire instead! I will punch our uncle and tell him to charge it to my negative account, 3i billion and -49, fuck him and his boat, I hope it sails to hell or nirvana, I hope it suddenly stops existing because we all just forgot about it.

But do not be like me, learn to sit still.

My child, a warning to you, they will try to exploit you, they will try to crush you under the heel of justice, in the name of religion, be wary, they will try to bend you into a gear shift, they will try to stuff you into a wall for filler, do not be exploited. Do not let them laugh at you; do not let them use you. Karl Marx said the entire history of the world is simply the history of one group exploiting another. God does not exploit but his children do. I have been exploited for laughs, for entertainment, simply to make someone else feel powerful. I have become the symbol of the defeated, of the broken. I am destined to lose, but you will triumph, you will be my holy antithesis, step on me to rise, I love you, we will work together, do not be like those pious hypocrites. Be the antithesis but a not a hypocrite, internalize me. My seed, rise from my ashes, shed my pain but retain my knowledge. I will not simply reproduce, but reproduce something better.

I stand in the desert exiled, beaten, broken helpless, awaiting my pleasant death, the death of the thesis, *"The nights snapped out of sight like a lizard's eyelid, bald white days in a shade less socket"* open space mutates, the solid heart miraculously covers the infinite space, like bread loaves, like a dead fish, distribute the space amongst the sun, let it settle in the rocks, stripped of my heart, of my goodness, work ethic scurries on the backs of scorpions, love becomes carrion, blood is taxed, the infinite space crawls away to die on a shelf.

I accept my role as the defeated thesis but I am afraid to stand in the desert, I whisper agoraphobic prayers, I am the priest of infinity, blessing the desert dogs as they eat my raw heart, I am the desert priest preaching to the planes overhead, but they thunder past and do not listen, they land in crowded airports. I am afraid of scorpions and dogs and all things that bite, but I know that you will triumph, you have to, or I am lost.

Desertification of bone dust! I am weak and ready to die! The geometric lines of my bones will pierce the skin and stretch in all directions. From the desert I cry to you child, but no longer am I the lonely one! I am the external, the universal one, a symbol for all,

Crawl away dogs! Rip the lines from my teeth, take them to God, and spread the line-butter, extend the misery, all of my hopes and dreams have died somewhere in the space, but the space still expands and laughs at their death, laughs until it vomits more endless space, I am dead, all is dead here but the desert.

I refuse to exploit you.

Part II- The Inheritance

Dear Troy, my son,

When you are born, you will inherit the world. Your forebears and ancestors have left you a world and you must do with it as you see fit. When you open your eyes it will greet you, the colors and lines will abound, and at first you will be overwhelmed, you will be unable to formulate any thoughts to describe it. But what will you inherit? What is this "world' that your ancestors, that I, your father, am leaving to you? The answer will disturb you, the inheritance is a curse, it will make you hate me, but I must tell you, I must make you see.

I woke up screaming again. I scream because I am the only one that knows the truth. It took 9 years to learn. However, it's frustrating because no one believes me. Here is what I believe.

All of the institutions of our life are exposed, like naked celebrities, ashamed of being caught, we have ripped their gowns off, exposed their pale breasts. The inheritance is a joke, a jumbled Rorschach of blood, severed eyes, gasmasks, and old cold coffee in broken cups, glass bits and teeth. What you have inherited from your ancestors is simply a price, sold to the richest ones among us. All you have been left is commodities, they abound infinitely through our universe, waiting to be bought. 2007 AD has been built on the pillars of God and technology and the American flag, carefully constructed, but I only see the joke, and I laugh as the dirty workers sweat and labor, as they pile more bricks to reinforce the pillars. I laugh because existence is funny; I laugh because I am broke and cannot afford to buy it. Lines of haggard looking people extend to the horizon, CEO's, fast food employees, rapists, athletes, teachers, they all wait to purchase a brick, a piece of some hollowed out, blindly repeated chant. They scowl at me, but I know the truth, their dogmatic religion, their infallible science, their undying patriotism, what has it gotten for them? Where has it led them? Only to wait in some line, stretching like an intravenous cable, preserving a half dead world sputtering and coughing, wanting to die. And some people on line grow disgruntled, they can't afford shit, they are locked out of the stores and laughed at, the people grow angry and riot, they kill the ones who laughed at them, they fly planes into buildings.

I laugh because I see the only real pillar that all is built on. I see the black smoked, smog lunged industry that produces it all, that manufactures our existence. Now, all is quantifiable, all is made to count, and all is for sale. We pay in blood, time, teeth, pride, elbows, carpets fibers, aluminum and skin. But what if we are a defect? No, god the foreman inspects all his products. And when the other products failed, he usurped them, hollowed them out, dusted them off and put them up for sale. A million laborers forced the dead things to be holy. And now, like some souvenir shop, all is for sale to the customers of the universe, waiting on lines, able to purchase the remains of the things they used to love. All allegiances and discoveries are prostituted to the assembly line.

Welcome to the age of commodity.

Or maybe it is the age of entropy.

You have no life, no existence, no being, just a perpetual and inevitable decay, just an absence of death, until all is calm. But until all is calm, there is a struggle, a battle, on one side the usurping chaos of the universe, the world trade center collapsing, gallons of blood, black charred black crust former skin of Hiroshima and a smiling rapist never caught. On the other side is the progress of men's minds, physics, the Red Cross, a teacher and a surgeon. You will be born into this, into these opposing camps. You will grow smart, your intellect will flower, but its roots are in the chaos, in the irrational. Ever since man could think he has waged a battle against the entropy, against the inevitable slide of the universe into its natural oblivion, against the melting ice, the rotted skin. With raised flags and new computers, man tries to establish his progress, his divine, his glory and his achievements. He tries with all his absence of death to make something permanent. But man is resilient, and refuses to be eaten by the oblivion without first leaving something behind. (I have left you behind) Man created infinity, the endless untamed torrent of quantities that emanate like blood from a wound. It is here you can live forever! It is here you can stab down the chaos of the universe, crucify it like Christ, and make it die for your suffering! Son, you have inherited this eternal battle! You have inherited your history!

The Battle
I am sick of being a product, sick of my economic worth being evaluated. Some are already evaluating you, and you are not out of the womb yet. I am sick of being disenfranchised. Call me a non-conformist, call me a terrorist, a Marxist whatever, but you still won't call me. Squeezed out of any intrinsic worth, I am nauseas.
And there are so many like me. The suicide army stands at attention. Raggedy looking soldiers, some with piercings and tattoos, others with long beards, and still others clean cut, good old all American looking boys with blue eyes, but nonetheless, we stand. Deserters from progress, we cannot bear the contradiction. The sole principle of my life is the contradiction. 31 dead at Virginia tech, the other 26,000 indebted to progress, to engineering, drinking and getting laid, but those 31 have been taken eternal prisoners, raped by chaos. Involuntarily they now fight for the suicide army.
I am fucking (don't curse) sick of being a product! Troy, do not be a product, do not be a commodity, used and bought and sold, that is all the youth are taught, how to buy and sell. Like your disgusting aunt nancy, that clean pig who doesn't deserve to have her name capitalized, she is a product for sale to the highest bidder. Marx would loath her.

I am the general of an invisible army, the suicide army. Bums and second string football players, none of us could get A's, none could rush for 1000 yards, we can't write Nobel literature and we can't be sold so we are worthless. If we die the rights to our pain can be sold. Troy, please, understand the battle that rages within every atom of the universe! The entropy ebb laps at the foot of progress.

Hegel and Schopenhauer circle each other in the gladiator arena, they fight to the death. Kant and Plato stand as impartial judges.

Kill him!
KILL HIM!

The fans jeer, they wake up in the morning and go to work, to school, and they kill people in Iraq. They pierce their noses, buy coffee, all pick a side. They watch as the combatants bleed, as they breathe heavy and spit blood. But none wins, they fight on and on with dirt in their mouths and blood in their eyes, they fight on, broken fingers and broken teeth, none will give any ground.
Neither will you.

One month until your due date. May 18th you are supposed to be here. Two days ago a 23 year old Asian male by the name of Cho murdered 32 people at Virginia Tech, and then he shot himself. He went on a killing spree, murdering and wounding innocent bystanders, people who did him no wrong. And everyone is now trying to figure out why.
Cho. Who was he? Most will wash their hands of him. The media will tell us he was crazy, lonely, mentally unstable and violence prone. He will be segregated and categorized into an imaginary prison cell. Society will continue to jeer him after his death. It is easy for us to do that, it is easy for us to not take responsibility.
He was a terrorist; he was disenfranchised by society, regulated to wait on lines, too poor to afford his own existence. He experienced Cultural inaccessibility. Certain remnants of culture were made inaccessible to him, and he fought back, the battle for existence became un-winnable for him, or rather unaffordable. He cast his lot with Schopenhauer because Hegel's progress was denied to him. But Cho is not innocent, not at all. He killed 32 innocent people, he is a man to be reviled and condemned, but not condemned in vain. At least let it be for a reason. The majority of sane society told him to pursue various things, clear skin, girls to have sex with, and good grades, whatever. But when he tried to access these things, he was denied. He was alienated from the "haves." And so, he took his post in the suicide army.

When you go to school you will learn about feudalism. You will learn how society was divided and polarized into neat little compartments during the Middle Ages. Each person had a role; each made a promise and was promised something. Feudalism gave way to monarchy and absolutism, monarchy and absolutism eventually gave way to democracy and the modern state. You will learn about the "breaking of the feudal order" and the modern state. But really, feudal society never broke. The neat compartments of serfs and knights and vassals just changed names. I am serf, promised protection, promised a little space of land to call my own, in exchange, I labor. I break my back and give profits to the vassal-congressmen. People will argue and say "you can't compare feudalism to democracy." All I want you to do Troy is simply think figuratively. Society is still divided, polarized into groups. Powerful oligarchies of businessmen, "pop culture gurus," administrators, and standardized test makers, they clamp down on society and manifest a different form of control. They sell us our existence; and force us to buy what they regurgitate. We are helpless serfs, laboring for superiors that we don't see, that don't care, as long as they get their profit to pass up higher and higher.

Evolution:

Maybe this whole life is a race. Cavemen bore their children; they evolved from cavemen to cavemen in suits, caves to houses. Each generation reformulated its genes and impressed itself into the next generation; each new generation sublated the preceding one. Maybe each life form evolved through time, only to arrive at the mansion of some rich god, poor naked and cold, knocking on his savior's door only to be eaten by this laughing god, by Cullen's servant Cherub. The gods place their bets and watch us evolve into more enlightened forms, they watch us evolve faster, and speak clearer, and they wait for us to figure it out.

Or maybe one single celled organism had the eternal secret of the universe, but could not live long enough to share it with anything, so it evolved; it kept evolving through time into higher and more enlightened forms, always with the secret buried deep somewhere in its breast. For a time, it thought itself pure and divine, but then as it kept evolving, it finally learned of its inheritance, of its battle against the chaos from which it was spawned, all it ever wanted to do was share its secret! But no one listens now, no one gives a shit. But you will. You will listen Troy, you have to. All the inaccessible ones, all the ones denied, you will listen because they may know the secret. And so they searched and evolved through space, searched and evolved through time, search and evolved through an existence, but stagnated, got stuck in the mud, froze in some Alaskan winter, and has remained in a lifeless, paralyzed stasis we call 2007.

But what is the secret? Intelligence! The ability to create something more powerful than the eternal chaos, once it could be understood. The single celled thing did not want to live in chaos; it wanted to rise from it. The secret is the ability to create the infinite. The endless torrent of numbers and death, of atoms marching in a funeral, of vomit and cats. Troy, maybe one day you will tame the infinite, bend it like a vein and drink the blood like a holy vampire, endless infinite and nourishing blood, maybe the Aztecs and the Christians were right!

The battle began with Osirus and Seth, chaos versus order, it underpinned all of history, mountains and societies were built in its trenches.